T0210670

Lecture Notes in Computer Science   9554

Commenced Publication in 1973
Founding and Former Series Editors:
Gerhard Goos, Juris Hartmanis, and Jan van Leeuwen

More information about this series at http://www.springer.com/series/7407

Stéphane Bonnevay · Pierrick Legrand
Nicolas Monmarché · Evelyne Lutton
Marc Schoenauer (Eds.)

Artificial Evolution

12th International Conference, Evolution Artificielle, EA 2015
Lyon, France, October 26–28, 2015
Revised Selected Papers

 Springer

Editors
Stéphane Bonnevay
Université Lyon 1
Lyon
France

Evelyne Lutton
INRA
Thirverval-Grignon
France

Pierrick Legrand
Université de Bordeaux
Bordeaux
France

Marc Schoenauer
Université Paris-Sud
Inria Saclay
Orsay
France

Nicolas Monmarché
Université de Tours
Tours
France

ISSN 0302-9743   ISSN 1611-3349 (electronic)
Lecture Notes in Computer Science
ISBN 978-3-319-31470-9   ISBN 978-3-319-31471-6   (eBook)
DOI 10.1007/978-3-319-31471-6

Library of Congress Control Number: 2016933471

LNCS Sublibrary: SL1 – Theoretical Computer Science and General Issues

Printed on acid-free paper

This Springer imprint is published by Springer Nature
The registered company is Springer International Publishing AG Switzerland

Preface

This LNCS volume includes the best papers presented at the 12th Biennial International Conference on Artificial Evolution, EA[1] 2015, held in Lyon (France). Previous EA editions took place in Bordeaux (2013), Angers (2011), Strasbourg (2009), Tours (2007), Lille (2005), Marseille (2003), Le Creusot (2001), Dunkerque (1999), Nimes (1997), Brest (1995), and Toulouse (1994).

Authors were invited to present original work relevant to artificial evolution, including, but not limited to: evolutionary computation, evolutionary optimization, co-evolution, artificial life, population dynamics, theory, algorithmics and modeling, implementations, application of evolutionary paradigms to the real world (industry, biosciences), other biologically inspired paradigms (swarm, artificial ants, artificial immune systems, cultural algorithms), memetic algorithms, multi-objective optimization, constraint handling, parallel algorithms, dynamic optimization, machine learning, and hybridization with other soft computing techniques.

Each submitted paper was reviewed by three members of the international Program Committee. Among the 31 submissions received, 18 papers were selected for oral presentation and eight other papers for poster presentation. For the previous editions, a selection of the best papers that were presented at the conference and further revised were published (see LNCS volumes 1063, 1363, 1829, 2310, 2936, 3871, 4926, 5975, 7401, and 8752). Exceptionally, for this edition, the high quality of the 18 papers selected for the oral presentation led us to include a revised version of all these papers in this volume of Springer's LNCS series.

We would like to express our sincere gratitude to our invited speakers: Darell Whitley and Guillaume Beslon.

The success of the conference resulted from the input of many people to whom I would like to express my appreciation: the members of Program Committee and the secondary reviewers for their careful reviews that ensure the quality of the selected papers and of the conference, the members of the Organizing Committee for their efficient work and dedication assisted by Véronique Deslandres and Eric Duchene, the members of the Steering Committee for their valuable assistance, and Aurélien Dumez for his support on the administration of the website.

I take this opportunity to thank the different partners whose financial and material support contributed to the organization of the conference: Polytech'Lyon, University Lyon 1, ERIC, LIRIS, and CNRS.

---

[1] As for previous editions of the conference, the EA acronym is based on the original French name "Evolution Artificielle."

Last but not least, I thank all the authors who submitted their research papers to the conference, and the authors of accepted papers who attended the conference to present their work. Thank you all.

February 2016

Stéphane Bonnevay

EA 2015 Chair
University of Lyon 1
ERIC Laboratory
France

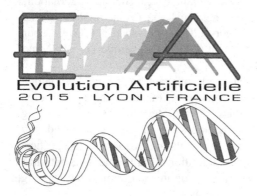

Évolution Artificielle 2015 – EA 2015

October 26–28, 2015
Lyon, France
12th International Conference on Artificial Evolution

Program Committee

Aguirre, Hernan	Shinshu University, Japan
Auger, Anne	Inria Saclay, France
Aupetit, Sébastien	University Francois Rabelais of Tours, France
Balev, Stefan	University of Le Havre, France
Bredeche, Nicolas	University Pierre et Marie Curie, France
Bonnevay, Stéphane	University of Lyon 1, France
Boumaza, Amine	University of Lorraine, France
Cagnoni, Stefano	University of Parma, Italy
Clergue, Manuel	University of the French West Indies, France
Collet, Pierre	University of Strasbourg, France
Daolio, Fabio	Shinshu University, Japan
Debbat, Fatima	University of Mascara, Algeria
Durand, Nicolas	ENAC, Toulouse, France
Dutot, Antoine	University of Le Havre, France
Ebner, Marc	University in Greifswald, Germany
Fonlupt, Cyril	University of the Littoral, Calais, France
Galvan, Edgar	Trinity College, Dublin, Ireland
Giacobini, Mario	Molecular Biotechnology Center, University of Turin, Italy
Hao, Jin-Kao	University of Angers, France
Idoumghar, Lhassane	University of Mulhouse, France
Jourdan, Laetitia	University of Lille, France
Langdon, Bill	University College, London, UK
Legrand, Pierrick	University of Bordeaux, France
Liefooghe, Arnaud	University of Lille 1, France
Lopez-Ibanez, Manuel	Université Libre de Bruxelles, Belgium
Louchet, Jean	Inria Saclay, France
Lutton, Evelyne	INRA, France
Marion-Poty, Virginie	University of the Littoral, France
Monmarché, Nicolas	University Francois Rabelais of Tours, France
Ochoa, Gabriela	Stirling University, Scotland, UK
Paquete, Luis	University of Coimbra, Portugal
Parkes, Andrew	University of Nottingham, UK
Pereira, Francisco	University of Coimbra, Portugal
Robilliard, Denis	University of the Littoral, France
Saubion, Frederic	University of Angers, France

Schoenauer, Marc	Inria Saclay, France
Siarry, Patrick	University of Paris-Est Creteil, France
Solnon, Christine	INSA Lyon, France
Stutzle, Thomas	IRIDIA, Brussels, Belgium
Talbi, El-Ghazali	Inria Lille, France
Teytaud, Olivier	Inria Saclay, France
Teytaud, Fabien	University of the Littoral, Calais, France
Tonda, Alberto	INRA, France
Urbano, Paulo	University of Lisbon, Portugal
Veerapen, Nadarajen	Stirling University, Scotland, UK
Verel, Sébastien	Université du Littoral Côte d'Opale, France
Z.-Flores, Emigdio	Instituto Tecnologico de Tijuana, Mexico

Steering Committee

Stéphane Bonnevay	Université Lyon 1, France
Pierre Collet	Université Louis Pasteur de Strasbourg, Strasbourg
Pierrick Legrand	Université de Bordeaux, France
Evelyne Lutton	INRA, France
Nicolas Monmarché	Université François Rabelais de Tours, France
Marc Schoenauer	Inria, France

Organizing Committee

Stéphane Bonnevay (*General Chair, Local Organization*)	Université Lyon 1, France
Véronique Deslandres (*Local Organization*)	Université Lyon 1, France
Eric Duchene (*Local Organization*)	Université Lyon 1, France
Aurelien Dumez (*Admin Web*)	Inria, France
Gérald Gavin (*Local Organization*)	Université Lyon 1, France
Laetitia Jourdan (*Publicity*)	Inria, France
Pierrick Legrand (*LNCS Publication*)	Université de Bordeaux, France
Sébastien Vérel (*Treasurer*)	Université du Littoral Côte d'Opale, France

Invited Speakers

Guillaume Beslon, Professor at the Computer Science Department of the National Institute of Applied Science in Lyon (France), which is part of the Université de Lyon. Member of the Laboratoire d'InfoRmatique en Image et Systèmes d'information (LIRIS, UMR 5205 CNRS); Head of the Inria Beagle Team, former director of Rhône-Alpes Institute of Complex Systems (IXXI).

Can Artificial Evolution Shed Light on Evolution of Complexity in Real Organisms?

Artificial evolution has a long a successful history in optimization. Yet, artificial evolution in a computer can also be used as a model of "real evolution." This field of research, known as "digital genetics" or "in silico experimental evolution," is rapidly growing and results accumulate rapidly. In this talk, I will present how in silico experimental evolution can be used to study the C-value paradox, an open question in biology for more  than 40 years. To this aim, I will present aevol, a simulation software developed by the LIRIS/Inria Beagle Team, and show how using such tools can shed new light, often counterintuitive, on this old question.

Darrell Whitley. Prof. Whitley is Chair of the Department of Computer Science at Colorado State University. From 1993 to 1997 Prof. Whitley served as Chair of the Governing Board of the International Society for Genetic Algorithms. In 1999 ISGA merged with the Genetic Programming community to form the International Society for Genetic and Evolutionary Computation. From 1997 to 2002 Prof. Whitley served as Editor-in-Chief for the journal *Evolutionary Computation* published by MIT Press. In 2005 ISGEC became a Special Interest Group (Sigevo) of ACM. In 2007 Prof. Whitley was elected Chair of Sigevo.

Blind No More: Deterministic Move and Recombination Operators for Evolutionary Algorithms

 For decades, most local search algorithms have relied on enumerating a neighborhood of solutions in order to locate improving moves. Evolutionary algorithms have similarly relied on random mutation and random recombination operators to generate new candidate solutions.

For k-bounded pseudo-Boolean optimization problems such as MAX-kSAT and NK-Landscapes, we have been able to prove it is possible to exactly identify improving bit flip moves in constant time under reasonable assumptions. Furthermore, this result can be generalized: We can also identify all improving moves within a Hamming radius r in constant time. This means that we no longer need to enumerate neighborhoods for local search, or to use random mutations to locate improving moves.

We can also prove that there exist deterministic forms of recombination that are also guaranteed to return the best possible offspring under reasonable assumptions. Given two parent solutions, the method identifies p subgraphs that partition the variable interactions of the parents. Given p subgraphs, recombination can be done in $O(n)$ time such that crossover returns the best solutions out of 2^p offspring. This form of "partition crossover" has been developed for both k-bounded pseudo-Boolean optimization problems as well as for the traveling salesman problem. Empirical results suggest that partition crossover is highly effective at accelerating search. We can now quickly generate globally optimal results for problems with $n = 100,000$.

Contents

The Multi-Funnel Structure of TSP Fitness Landscapes: A Visual Exploration

Gabriela Ochoa[1]([✉]), Nadarajen Veerapen[1], Darrell Whitley[2],
and Edmund K. Burke[1]

[1] Computing Science and Mathematics, University of Stirling,
Stirling, Scotland, UK
gabriela.ochoa@cs.stir.ac.uk
[2] Department of Computer Science, Colorado State University, Fort Collins, USA

Abstract. We use the Local Optima Network model to study the structure of symmetric TSP fitness landscapes. The 'big-valley' hypothesis holds that for TSP and other combinatorial problems, local optima are not randomly distributed, instead they tend to be clustered around the global optimum. However, a recent study has observed that, for solutions close in evaluation to the global optimum, this structure breaks down into multiple valleys, forming what has been called 'multiple funnels'. The multiple funnel concept implies that local optima are organised into clusters, so that a particular local optimum largely belongs to a particular funnel. Our study is the first to extract and visualise local optima networks for TSP and is based on a sampling methodology relying on the Chained Lin-Kernighan algorithm. We confirm the existence of multiple funnels on two selected TSP instances, finding additional funnels in a previously studied instance. Our results suggests that transitions among funnels are possible using operators such as 'double-bridge'. However, for consistently escaping sub-optimal funnels, more robust escaping mechanisms are required.

1 Introduction

The structure of combinatorial fitness landscapes is known to impact the performance of heuristic search algorithms. Features such as the number and distribution of local optima and their basins of attraction are among the most studied. The relationship among local optima for the symmetric Traveling Salesman Problem (TSP) under the standard 2-change neighbourhood was first analysed in [4], where a *globally convex* structure was discovered. The global optimum was found to be 'central' to all other local optima conforming a 'big-valley' structure. This is interpreted as a landscape where many local optima exists, but they are easy to escape and the gradient, when viewed at a coarse level, leads to the global optimum (Fig. 1). However, a more recent study has found that the big valley structure breaks down when considering solutions near in evaluation to the global optimum [7]. The big-valley separates into multiple valleys, conforming what has been called 'multiple funnels' in the study of energy surfaces

© Springer International Publishing Switzerland 2016
S. Bonnevay et al. (Eds.): EA 2015, LNCS 9554, pp. 1–13, 2016.
DOI: 10.1007/978-3-319-31471-6_1

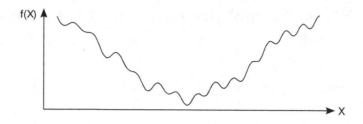

Fig. 1. Depiction of the 'big-valley' structure.

in chemical-physics [19]. The multi-funnel concept implies that local optima are organised into clusters, so that a particular local optimum largely belongs to a particular funnel. The appearance of multiple funnels explains why certain iterated local search heuristics can quickly find high-quality solutions, but fail to consistently find the global optimum. In a series of studies, Whitley et al. [7,20,21] have proposed a crossover operator (Partition Crossover), which has demonstrated the ability to escape funnels at evaluations close to the global optimum. A similar recombination operator [12] is used by Helsgaun [8] in the so called LKH-solver.

This article uses the Local Optima Network (LON) model [14–16,18] in order to explore in more detail the structure of TSP landscapes near the global optimum. Local optima networks compress the whole search spaces into a graph having as vertices the local optima, and as edges transitions among them according to a given search operator. This network-based model brings the tools from the *new science of networks* [13] (e.g., metrics and visualisation) to the study of fitness landscapes in combinatorial optimisation.

Our study considers Chained Lin-Kernighan (Chained-LK), one of the best performing heuristic algorithms for TSP [2,11]. Chained LK is an iterated local search approach combining the variable depth local search of Lin and Kernighan (LK-search) [10] with the *double-bridge* move [11] (a form of 4-change, depicted in Fig. 2b) as the perturbation or 'kick' operator. Therefore, the proposed LON model considers local minima according to LK-search, and transitions among them according to the double-bridge move. Our goal is to gain a deeper understanding of the multi-funnel structure of the TSP under Chained-LK, which will help in selecting and designing stronger escape mechanisms (such as Partition Crossover [20,21]) to avoid being trapped in a sub-optimal funnel. The main contributions of this article are the following:

1. First study of local optima networks for TSP, including their sampling and analysis.
2. Definition of the DLON model (distance local optima networks) and adaptation of the escape edges model (ELON) to TSP.
3. Network visualisation of the multi-funnel structure of TSP fitness landscapes.

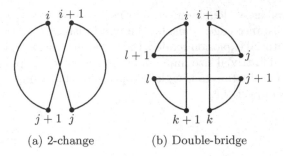

(a) 2-change   (b) Double-bridge

Fig. 2. Illustration of tours obtained after 2-change or double-bridge move.

2 Local Optima Networks for TSP

For a TSP instance with n cities, the search space is the set of permutations of the n cities. The number of tours, which equals the number of permutations, is factorial in n. The fitness function f is given by the length of the tour, which is to be minimised. Before presenting formal definitions in Sect. 2.1, we briefly describe the following notions relevant to our model.

LK-search: The well-known Lin-Kernighan heuristic is a powerful local search algorithm. It is based on the idea of k-change moves: take the current tour and remove k different links from it, which are then reconnected in a new way to achieve a legal tour. A tour is considered to be 'k-opt' if no k-change exists which decreases its length. Figure 2a illustrates a 2-change move. LK applies 2, 3 and higher-order k-changes. The order of a change is not predetermined, rather k is increased until a stopping criterion is met. Thus many kinds of k-changes and all 3-changes are included. There are many ways to choose the stopping criteria and the best implementations are rather involved. We use here the implementation available in the Concorde software package [1], which uses *do not look* bits and candidate lists.

Double-Bridge move: Proposed by Martin et al. [11] as the 'kick' mechanism in the Chained-LK heuristic, the double-bridge move (drawn in Fig. 2b) is a type of 4-change. It consists of two improper 2-changes, each of which is a 'bridge' (i.e., it takes a legal, connected tour into two disconnected parts). The combination of both bridges, must then be chosen as to produce a legal final tour.

Bond distance: Measures the difference between two tours t_1 and t_2 according to the number of edges or 'bonds' that differ in both tours. Specifically, $b(t_1, t_2)$ is equal to n minus the number of edges that are present in both t_1 and t_2 disregarding edge direction [4].

Our approach requires defining and extracting local optima networks for TSP instances. To construct the networks, we need to define their nodes and edges. Nodes will be local optima according to LK-search, and two types of weighted edges are considered: escape edges and distance edges. The escape edges are based on the number of double-bridge moves required to escape from a local optimum, while distance edges consider the bond distance between solutions.

Since combinatorial explosion renders the full enumeration of local optima for TSP instances of non-trivial size impossible, we resort to sampling local optima which are close in evaluation to the global optimum. The sampling procedure is further described in Sect. 2.2.

2.1   Definitions

Definition 1. A *funnel floor* solution is a high quality local optimum that is conjectured to be at the bottom of a funnel. Indeed, they were called funnel bottom solutions in [7], and are generated running Chained-LK for a large enough number of iterations. The set of funnel floor solutions is denoted by F.

Definition 2. A *funnel basin* solution is a local optimum within a funnel. Each funnel basin solution is obtained by first locating a funnel floor, and then escaping from the funnel floor in order to discover a nearby local optimum. In this article, this is done using a random walk with double-bridge followed by improvement using LK-search. The set of local optima defining the funnel basins is denoted by B. Specifically, for some $x \in F$, $y \in B_x \subseteq B$ if it can be obtained from x after a sequence of length d of double-bridge moves followed by LK-search. Since after a double-bridge followed by LK-search the local optimum obtained y can be equal to the starting point x, the length d of the random walk is incremented until $y \neq x$.

The set of local optima, L, is the union of the funnel floors and local optima that define the funnel basins, $L = F \cup B$.

Definition 3. An *escape edge* is a weighted edge from a funnel floor to a local optimum. Specifically, there is an edge $e_{x,y}$ of weight d between the funnel floor point $x \in F$ and the local optima $y \in B$ if y can be obtained from x after a sequence of length d of double-bridge moves followed by LK-search. No self-loops are considered. The set of escape edges is denoted by E_{esc}.

Definition 4. A *distance edge* is a weighted edge, according to the bond distance, between any two local optima. Specifically, there is an edge $e_{x,y}$ of weight d between local optima x and $y \in L$ if the bond distance $b(x, y) = d$. The set of distance edges between any two local optima in L is denoted by E_{dist}.

Definition 5. The *Escape Local Optima Network (ELON)* is the graph $ELON = (L, E_{esc})$ where nodes are the local optima L, and edges E_{esc} are the escape edges.

Definition 6. The *Distance Local Optima Network (DLON)* is the graph $DLON = (L, E_{dist})$ where nodes are the local optima L, and edges E_{dist} are the distance edges.

Data: I, a TSP instance
Result: F, the set of tours on the funnel floors
          B, the set containing the escape tours from sampled funnel floors
$F \leftarrow \emptyset$;
for $i \leftarrow 1$ **to** $10,000$ **do**
   | $x \leftarrow$ chainedLK$(I, \text{stallcount} = 10,000)$;
   | **if** $x \notin F$ **then**
   | | $F \leftarrow F \cup \{x\}$;
   | **end**
end
$S \leftarrow$ mostFrequentSolutionForEachFitnessLevel(F);
$B \leftarrow \emptyset$;
for $v_0 \in S$ **do**
   | $B_{v_0} \leftarrow \emptyset$;
   | **for** $j \leftarrow 1$ **to** $1,000$ **do**
   | | $i \leftarrow 0$;
   | | **repeat**
   | | | $i \leftarrow i + 1$;
   | | | $v_i \leftarrow$ randomDoubleBridgeMove(v_{i-1});
   | | | $v' \leftarrow$ LK(v_i);
   | | **until** $v' \neq v_0$;
   | | $B_{v_0} \leftarrow B_{v_0} \cup \{v'\}$;
   | **end**
   | $B \leftarrow B \cup B_{v_0}$;
end

Algorithm 1. TSP local optima network sampling procedure

2.2 Sampling Methodology

We apply a sampling strategy similar to that used by Hains et al. [7] where two stages are considered. This process also resembles the one used by Iclanzan et al. [9] to sample the landscape of Quadratic Assignment Problem instances. In the first stage, local optima of very good quality are identified which define the funnel floors (set F defined in Sect. 2.1). In the second stage, random walks are generated to escape these local optima in order to determine the funnels' basins (set B defined in Sect. 2.1). These approaches are detailed below and through pseudocode in Algorithm 1.

The funnel floor solutions are tours obtained when Chained-LK stalls. In practice, we determine stalling to occur when fitness does not improve for 10,000 consecutive iterations of Chained-LK. This procedure is itself repeated 10,000 times from a randomly generated initial tour and the unique tours produced are saved in F, the set of funnel floor solutions. This procedure corresponds to the first loop in Algorithm 1.

To determine a funnel's basin, we identify a start point in its floor, let us call it v_0, and follow a random walk using a sequence of double bridge perturbations. More precisely, at each step i of the random walk, a random move is performed on v_{i-1}, producing a tour v_i. An LK-search is then applied to v_i to produce a locally optimal tour v'. If v' is different from v_0, then we have escaped from the basin of attraction of v_0. The random walk is stopped and its length i is the escape distance. Tour v' is saved in B_{v_0}, the set of tours having escaped from v_0. This escape procedure is repeated 1,000 times.

When there are many tours on the funnel floors, it is impractical to try to escape from all of them. When Hains et al. [7] computed the funnels floors from 1,000 Chained-LK applications, they found that tours with the same fitness level formed a connected component under 2-change. These could thus be considered to form a plateau and they, therefore, randomly chose one tour to escape from out of each plateau.

In our case, having performed 10,000 Chained-LK applications, we find many more tours on the funnel floors and, furthermore, they are not all on 2-change plateaus. Our approach selects the most frequently occurring solution within each fitness level as a starting solution. Ties are broken at random.

3   Results

Our study considers two 'milestone' TSP instances: *lin318* and *att532* (as named in TSPLIB [17], also listed in Table 1.5 from [3]). They are composed of 318 and 532 cities, and were first solved to optimality in 1980 and 1987, respectively. The lin318 instance is a circuit board drilling example (i.e., it models the routing of a numerically controlled drilling machine efficiently through a set of hole positions), and was presented by Lin and Kernighan in their seminal paper [10]. It remained the largest TSP instance solved to optimality for a span of seven years in the 1980s. The att532 instance is comprised of pseudo-Euclidean coordinates that go through the 532 largest cities of the USA. It is very well known given the difficulty that the distances to the next node are very short at the east coast, whereas in other regions of the USA they are very long.

Results are discussed in the following two subsections. Section 3.1 analyses the sampled local optima and the bond and escape distances among them. Section 3.2 visualises the escape and distance local optima networks.

3.1   Local Optima and Distances

For instance lin318, 4 unique funnel floor solutions were identified, each with a different fitness level (Table 1). The global optimum was found in the overwhelming majority, 96 %, of cases. The other funnel floor solutions' fitness is within 0.32 % of the global optimum.

When considering att532, 47 unique funnel floor solutions were identified, distributed among 8 different fitness levels (Table 2). This is in contrast to the 20 unique solutions and 4 different fitness levels found by Hains et al. [7]. A closer look at the data reveals that these 4 fitness levels amount to the most frequent fitness levels in our data, comprising 99 % of the solutions found. The seldom found solutions are therefore a result of carrying out a greater number of Chained-LK searches to sample solutions close to the global optimum.

Table 1. lin318 summary data

	All Sols	Fitness Levels			
		42029	42143	42155	42163
Unique Solutions	4	1	1	1	1
Fitness Level Freq. (%)		96.02	3.59	0.09	0.30
Colour of funnel in figures		■	■	■	■
Symbol in Fig. 4a		○	△	+	×

Table 2. att532 summary data

	All Sols	Fitness Levels							
		27686	27693	27703	27704	27705	27706	27708	27715
Unique Solutions	47	2	1	8	8	13	8	5	2
Fitness Level Freq. (%)		41.78	0.04	33.17	0.65	20.69	3.58	0.07	0.02
Start Point Freq. (%)		21.35	0.04	5.80	0.16	4.64	0.57	0.03	0.01
Colour of funnel in figures		■	■	■	■	■	■	■	■
Symbol in Fig. 4b		○	△	+	×	◇	▽	⊠	✳

The two globally optimal solutions account for only 42 % of all funnel floor solutions found but all the fitnesses are within 0.10 % of the global optimum. As previously mentioned, for att532, the starting points we try to escape from are the most frequent funnel floor solution within each fitness level. These make up 33 % of the solutions found.

The pairwise bond distances between the starting points for both instances are given in Fig. 3. In most cases, the pairwise distance between any two solutions is non-trivial. For example, the bond distance between the first two best solutions for lin318 is 37.

For att532, the smallest bond distance between start points is only 16. This seems to be a bridgeable distance with a small number of double-bridge moves. The starting point with fitness 27,693 only represents 0.04 % of funnel floor solutions. It is at distance 16 from the start point with fitness 27,686 that constitutes 21 % of solutions found. These numbers suggest that there is a reasonable way to move between these funnels, which explains why so few solutions with fitness 27,693 are found. This is corroborated by the local optima networks visualised in Sect. 3.2.

To analyse the fitness distribution of local optima within funnels, let us consider Fig. 4. Dot plots of fitness versus bond distance to the global optimum are presented for both instances. In addition, kernel density estimation distributions of points are provided.

Here our results match those of Hains et al. [7]. Firstly, local optima within a funnel are correlated in fitness and distance to their own respective starting

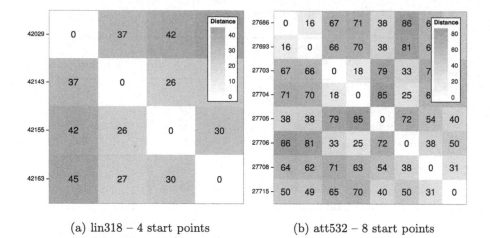

(a) lin318 − 4 start points   (b) att532 − 8 start points

Fig. 3. Pairwise distances between funnel floor solutions for instances lin318 and att532. Fitness levels are indicated on the left of each plot. In (a), instance lin318 has a single solution per fitness level. In (b), the most frequent solution is selected for each fitness level of att532.

point. Secondly, there is little correlation between fitness of local optima near the global optimum and their distance to it. However, for att532, the great majority of the local optima observed by Hains et al. when using double-bridge were below the 27,750 fitness level and a plot similar to ours was only obtained when using 2-change instead of double-bridge. They therefore concluded that double-bridge exacerbates the multi-funnel structure. We found instead that, when comparing the two escape operators, it is 2-change that exacerbates the multi-funnel structure. In other words, it is harder to escape funnels using 2-change as compared to double-bridge.

Figure 5 gives the escape and pairwise bond distance distributions for both instances. With a mean and mode of 1 for the escape distance, we can see that the double-bridge move is highly effective in escaping from the starting points.

For bond distances, the distribution for all edges differs from the distribution considering only edges between a start point and the solutions it escaped to. For lin318, when considering all start points, the distribution roughly resembles a step function with 2 steps which then quickly tapers off. The same distribution can be observed when considering each start point separately (not shown here). For att532, the bond distance distributions when considering a single start point to the local optima within the funnel appear to be bimodal (not shown here) or similar to the distribution when considering all start points. We intend to look more closely at distributions within individual funnels in future work.

3.2 Local Optima Networks

The two local optima networks models, using escape and bond distance edges, were extracted and visualised for the two selected TSP instances. Both models

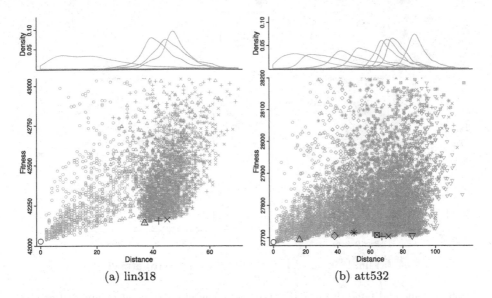

(a) lin318   (b) att532

Fig. 4. Dot plots and corresponding density distribution plots of the local optima generated when escaping from funnel floors. Bond distance is computed w.r.t. to the global optimum, or the most frequent of the two global optima in the case of att532. The range of fitness values displayed is chosen to encompass at least 95 % of points. Start points are indicated by a black symbol (Colour figure online).

clearly suggest a multi-cluster (multi-funnel) structure (see Fig. 6 explained below). The escape edges give a network view of the search process by Chained-LK, while the bond distance model is more general and illustrates the distributions of local optima which are close in distance.

At the heart of network visualisation is the graph layout. We use here the Fruchterman and Reingold's method [6] provided by the igraph package [5] for the R statistical language. The method is based on exploiting analogies between the relational structure in graphs and the forces among elements in physical systems. Specifically, considering attractive and repulsive forces by associating vertices with balls and edges with springs. The heuristic is concerned with drawing graphs according to some generally accepted aesthetic criteria such as (a) distribute the vertices evenly in the frame, (b) minimise edge crossings, (c) make edge lengths uniform, and (d) reflect inherent symmetry [6].

Figure 6 visualises the two network models (escape and distance edges) on the two studied instances. In order the make the picture manageable in size, sub-graphs of the whole sampled networks were selected for visualisation. The sub-graphs include all the funnel floor solutions (drawn as squares), and all the solutions that we call *frontier* nodes (drawn in black). These frontier nodes are those that can be attained from more than one funnel start point by the escaping mechanisms (i.e., a sequence of double bridge moves followed by LK-Search). The colour of the remaining nodes indicates the funnel (fitness level) membership (as indicated in Tables 1 and 2 for lin318 and att532, respectively) with the red colour

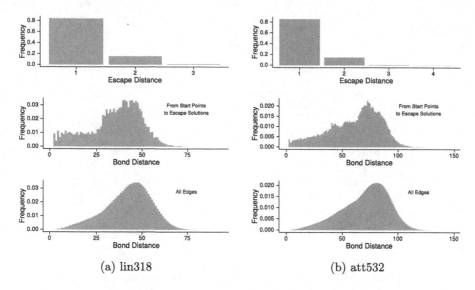

Fig. 5. Escape and bond distance distributions. The most frequent escape distance is 1. The maximum escape distance is 4 on att532, but occurred only once.

identifying the funnel of the global optimum. For the lin318 instance, 10 % of the funnel basin points were selected for visualisation. This percentage was 5 % for the larger att532 instance. All the escape edges are visualised, with darker grey indicating edges with escape distance 1. Visualising all bond distance edges is not feasible, so we set a threshold of 1/10 of the maximum distance to the global optimum in the sampled points (i.e., there is an edge if the distance between nodes is below the given threshold). This threshold was a distance of 9 for lin318 and 14 att532. Again the darker grey identifies edges with the minimum distance.

The multi-funnel structure can be visualised in the network plots in Fig. 6, which separate in clearly defined clusters of solutions. The lin318 instance features 4 clusters, while att532 has 8 clusters. The clusters are more clearly defined for the escape edges, but interestingly, the same overall structure appears for the distance edges. It is interesting to observe that some points (drawn in black) 'belong' to more than one funnel. That is, they can be reached from more than one funnel floor by double-bridge moves followed by LK-search. Therefore, it is possible for Chained-LK to escape some funnels, but it seems difficult for it to consistently escape from all funnels.

An interpretation of the effectiveness of Chained-LK may be obtained when considering the local optima networks together with the fitness levels of the start points of each funnel, their frequency when sampling the funnel floors and the pairwise bond distance between start points.

For lin318, the two connected funnels are the ones whose start points have fitness 42,143 and 42,155 and were sampled 3.59 % and 0.09 % of the time respectively. They are also the two closest start points for lin318, with a distance of 26. For att532, as was observed in Sect. 3.1, start points with fitness 27,686 and 27,693 are at a distance of 16 and constitute 21.35 % and 0.04 % of sampled

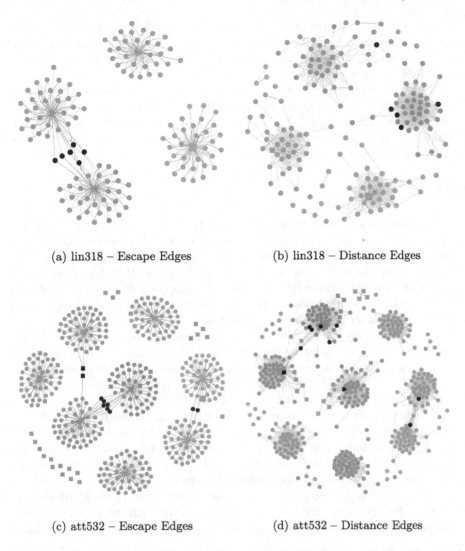

(a) lin318 – Escape Edges

(b) lin318 – Distance Edges

(c) att532 – Escape Edges

(d) att532 – Distance Edges

Fig. 6. Visualisation of Local Optima Networks for lin318 (top) and att532 (bottom). Both networks models, using escape and distance edges, are visualised. Nodes are local optima and edges represent escape or distance edges (with a set threshold), respectively. Square nodes represent solutions that belong to the funnel floors, while circle nodes to funnel basins. The larger square nodes (in red) are the global optima. Colours identify the different funnels (or fitness levels) as indicated in Tables 1 and 2. The black nodes are 'frontier' points, i.e., points that can be reached from more than one funnel. (Color figure online)

funnel floors. They are at a bond distance of 16 to each other and their corresponding funnels are linked in the local optima networks. The start point with fitness 27,703 (5.80 %) is connected to the start point with fitness 27,704 (0.16 % and distance 18). While these three observations are not sufficient to draw broad conclusions, an initial interpretation is that 'close enough' start points exhibit funnels that are linked to each other. Furthermore, when two funnels are connected, it is highly probable that the search will end up in the funnel with the funnel floor with better fitness.

The start point with fitness 27,703 is also connected to the one with fitness 27,706 (0.57 % and distance 33), but through two other floor solutions (indicated by black squares in the figure) that were not used as start points. These two floor solutions are also of fitness 27,703 and 27,706 and are only at a bond distance of two from the start point with the same fitness.

4   Conclusions

We have implemented a sampling procedure to extract local optima networks for TSP instances. In particular, we studied the search space structure close to the global optimum and confirmed the existence of multiple funnels. Our study is the first to analyse local optima networks for TSP and provide a clear visualisation of its multi-funnel structure. The proposed distance local optima network model is a contribution of this article, which may find easy application in other combinatorial optimisation problems.

Our analysis considered the well-known Chained-LK heuristic as implemented in the Concorde software package. Chained-LK is an iterated local search approach combining LK-search with double-bridge as the perturbation or escape operator. On two selected TSP instances, we found that while some funnels are directly connected to other funnels via double-bridge escape moves, most of them are not. This gives a visual insight of why Chained-LK produces suboptimal solutions in some runs, and justify the multiple restarts used in the default Concorde implementation. We hypothesise that when Chained-LK produces sub-optimal solutions, it is because it gets trapped in a sub-optimal funnel and the double-bridge escape mechanism, while generally efficient to escape local optima, is not strong enough to escape some funnels. Future work will explore alternative funnel-escape mechanisms such as the recently proposed Partition Crossover [20,21], and will study Tunneling Crossover Networks for TSP [14].

Acknowledgements. This work was supported by the UK's Engineering and Physical Sciences Research Council [grant number EP/J017515/1].

Data Access. All data generated during this research are openly available from the Zenodo repository (http://doi.org/10.5281/zenodo.20732).

References

1. Applegate, D., Bixby, R., Chvátal, V., Cook, W.: Concorde TSP solver (2003). http://www.math.uwaterloo.ca/tsp/concorde.html
2. Applegate, D., Cook, W., Rohe, A.: Chained Lin-Kernighan for large traveling salesman problems. INFORMS J. Comput. **15**, 82–92 (2003)
3. Applegate, D.L., Bixby, R.E., Chvátal, V., Cook, W.J.: The Traveling Salesman Problem: A Computational Study. Princeton University Press, Princeton (2007)

4. Boese, K.D., Kahng, A.B., Muddu, S.: A new adaptive multi-start technique for combinatorial global optimizations. Oper. Res. Lett. **16**, 101–113 (1994)
5. Csardi, G., Nepusz, T.: The igraph software package for complex network research. InterJournal Complex System, 1695 (2006)
6. Fruchterman, T.M.J., Reingold, E.M.: Graph drawing by force-directed placement. Softw. Pract. Exper. **21**(11), 1129–1164 (1991)
7. Hains, D.R., Whitley, L.D., Howe, A.E.: Revisiting the big valley search space structure in the TSP. J. Oper. Res. Soc. **62**(2), 305–312 (2011)
8. Helsgaun, K.: An effective implementation of the LinKernighan traveling salesman heuristic. Eur. J. Oper. Res. **126**(1), 106–130 (2000)
9. Iclanzan, D., Daolio, F., Tomassini, M.: Data-driven local optima network characterization of QAPLIB instances. In: Proceedings of the 2014 Conference on Genetic and Evolutionary Computation, GECCO 2014, pp. 453–460. ACM, New York (2014)
10. Lin, S., Kernighan, B.W.: An effective heuristic algorithm for the traveling-salesman problem. Oper. Res. **21**, 498–516 (1973)
11. Martin, O., Otto, S.W., Felten, E.W.: Large-step Markov chains for the traveling salesman problem. Complex Syst. **5**, 299–326 (1991)
12. Möbius, A., Freisleben, B., Merz, P., Schreiber, M.: Combinatorial optimization by iterative partial transcription. Phys. Rev. E **59**(4), 4667–4674 (1999)
13. Newman, M.E.J.: Networks: An Introduction. Oxford University Press, Oxford (2010)
14. Ochoa, G., Chicano, F., Tinos, R., Whitley, D.: Tunnelling crossover networks. In: Proceedings of the Genetic and Evolutionary Computation Conference (GECCO), pp. 449–456. ACM (2015)
15. Ochoa, G., Tomassini, M., Verel, S., Darabos, C.: A study of NK landscapes' basins and local optima networks. In: Proceedings of the Genetic and Evolutionary Computation Conference (GECCO), pp. 555–562. ACM (2008)
16. Ochoa, G., Verel, S., Daolio, F., Tomassini, M.: Local optima networks: a new model of combinatorial fitness landscapes. In: Richter, H., Engelbrecht, A. (eds.) Recent Advances in the Theory and Application of Fitness Landscapes. ECC, vol. 6, pp. 233–262. Springer, Heidelberg (2014)
17. Reinelt, G.: TSPLIB-a traveling salesman problem library. ORSA J. Comput. **3**(4), 376–384 (1991). http://www.iwr.uni-heidelberg.de/groups/comopt/software/TSPLIB95/
18. Verel, S., Ochoa, G., Tomassini, M.: Local optima networks of NK landscapes with neutrality. IEEE Trans. Evol. Comput. **15**(6), 783–797 (2011)
19. Wales, D.J., Miller, M.A., Walsh, T.R.: Archetypal energy landscapes. Nature **394**, 758–760 (1998)
20. Whitley, D., Hains, D., Howe, A.: Tunneling between optima: partition crossover for the traveling salesman problem. In: Proceedings Genetic and Evolutionary Computation Conference, GECCO 2009, pp. 915–922. ACM, New York (2009)
21. Whitley, D., Hains, D., Howe, A.: A hybrid genetic algorithm for the traveling salesman problem using generalized partition crossover. In: Schaefer, R., Cotta, C., Kołodziej, J., Rudolph, G. (eds.) PPSN XI. LNCS, vol. 6238, pp. 566–575. Springer, Heidelberg (2010)

Approaches for Many-Objective Optimization: Analysis and Comparison on MNK-Landscapes

Hernán Aguirre[1]([✉]), Saúl Zapotecas[1], Arnaud Liefooghe[2,3],
Sébastien Verel[4], and Kiyoshi Tanaka[1]

[1] Faculty of Engineering, Shinshu University, 4-17-1 Wakasato,
Nagano 380-8553, Japan
{ahernan,ktanaka}@shinshu-u.ac.jp
[2] Université Lille, CNRS, Centrale Lille, UMR 9189 - CRIStAL,
59000 Lille, France
arnaud.liefooghe@univ-lille1.fr
[3] Inria Lille - Nord Europe, 59650 Villeneuve d'Ascq, France
[4] Université du Littoral Côte d'Opale, LISIC, 62228 Calais, France
verel@univ-littoral.fr

Abstract. This work analyses the behavior and compares the performance of MOEA/D, IBEA using the binary additive ε and the hypervolume difference indicators, and AεSεH as representative algorithms of decomposition, indicators, and ε-dominance based approaches for many-objective optimization. We use small MNK-landscapes to trace the dynamics of the algorithms generating high-resolution approximations of the Pareto optimal set. Also, we use large MNK-landscapes to analyze their scalability to larger search spaces.

1 Introduction

Recently, several algorithms are being proposed for many-objective optimization. Preferred approaches to implement selection in many-objective optimization are decomposition, performance indicators, and relaxations of Pareto dominance.

Decomposition based algorithms [1,2] break down the many-objective problem into a large number of single-objective problems using scalarizing functions. The single objective problems are then solved concurrently. The scalarizing functions are usually defined in advance and remain fixed during the search. To create a set of scalarizing functions we assume a distribution of the Pareto optimal front and the algorithm aims to find good solutions that match our assumptions on distribution. Indicator based algorithms use a performance indicator function to assess the quality of a set of solutions. These algorithms optimize a single-objective function aiming to find the best subset of Pareto non-dominated solutions according to the performance indicator [3–5]. Popular indicators are additive ε, hypervolume, and R2. Relaxations of Pareto dominance modify the dominance relation to discern between initially incomparable solutions. One effective approach to relax Pareto dominance is ε-dominance [6]. ε-dominance based algorithms expand the area of dominance of some non-dominated solutions

© Springer International Publishing Switzerland 2016
S. Bonnevay et al. (Eds.): EA 2015, LNCS 9554, pp. 14–28, 2016.
DOI: 10.1007/978-3-319-31471-6_2

using a mapping function that depends on a parameter ε. These algorithms use ε-dominance principles to update the archive [7] or sample the instantaneous population [8] in order to keep a subset of solutions spaced with the resolution induced by the ε mapping function. These three different approaches have led to many-objective algorithms that perform significantly better than conventional multi-objective algorithms on many-objective problems. However, there is not much work comparing them in a rigorous way and their dynamics solving many-objective problems is not yet fully understood.

This work analyses the behavior of representative algorithms that implement the above three main approaches for selection, namely the decomposition based MOEA/D, the indicator based IBEA using the binary additive ε-indicator and the binary hypervolume difference-indicator, and the ε-dominance based AεSεH. As reference, it also includes results by NSGA-II [9]. First, we use MNK-landscapes with 20 bits to trace the dynamics of the algorithms finding new optimal solutions and compare their performance generating high-resolution approximations of the Pareto optimal set. Then we use MNK-landscapes with 100 bits and analyze their scalability to larger search spaces. This work reveals important strengths and limitations of these algorithms for many-objective optimization, explaining their behavior and performance when convergence and diversity of the approximation is considered.

2 Algorithms

2.1 MOEA/D (Multiobjective EA Based on Decomposition)

MOEA/D [2] is a decomposition-based EMO algorithm that seeks high-quality solutions in multiple regions of the objective space by decomposing the original (multi-objective) problem into a number of scalarizing (single-objective) sub-problems. MOEA/D defines a neighboring relation among sub-problems, based on the assumption that a given sub-problem is likely to benefit from the current solutions maintained in the corresponding neighboring sub-problems. More particularly, let μ be the user-defined number of sub-problems. A set $(\lambda^1, \ldots, \lambda^i, \ldots, \lambda^\mu)$ of uniformly-distributed weighting coefficient vectors defines the scalarizing sub-problems, and a population $\mathcal{P} = (x^1, \ldots, x^i, \ldots, x^\mu)$ is maintained such that each individual x^i maps to the current solution of the corresponding sub-problem defined by λ^i. In addition, a set of neighbors $Neig(i)$ is defined by considering the T closest weighting coefficient vectors for each sub-problem i (including itself), $i \in \{1, \ldots, \mu\}$. At each iteration, the population evolves with respect to a given sub-problem i. Two solutions are selected at random from $Neig(i)$ and an offspring is produced by means of crossover and mutation operators. Then, for each sub-problem $j \in Neig(i)$, the offspring x is used to replace the current solution x^j if there is an improvement in terms of the defined scalarizing function. The algorithm iterates over sub-problems until a stopping condition is satisfied.

Different scalarizing functions can be used within MOEA/D. In this paper, we use the weighted Chebyshev metric defined below.

$$g(x, \lambda) = \max_{i \in \{1, \ldots, m\}} \lambda_i \cdot \left| z_i^\star - f_i(x) \right| \tag{1}$$

such that x belongs to the solution space, λ is a weighting coefficient vector and $z^\star$ is a reference point.

2.2   IBEA (Indicator-Based Evolutionary Algorithm)

IBEA [3] tries to introduce a total order between solutions by means of an arbitrary binary quality indicator I. The fitness assignment scheme of IBEA is based on a pairwise comparison of solutions in a population with respect to indicator I. Each individual x is assigned a fitness value measuring the "loss in quality" in the population P if x was removed from it as follows

$$\text{Fitness}(x) = \sum_{x' \in P \setminus \{x\}} (-e^{-I(x',x)/\kappa}), \tag{2}$$

where $\kappa > 0$ is a user-defined scaling factor. Survival selection is based on an elitist strategy that combines the current population $\mathcal{P}_t$ with its offspring $\mathcal{Q}_t$, iteratively deletes worst solutions until the required population size is reached, and assigns the resulting population to $\mathcal{P}_{(t+1)}$. Here, each time a solution is deleted the fitness values of the remaining individuals are updated. Parent selection for reproduction consists of binary tournaments between randomly chosen individuals using their fitness to decide the winners.

Several indicators can be used within IBEA. Here we choose to use the binary additive ϵ-indicator ($I_{\epsilon+}$) and the binary hypervolume difference-indicator (I_{HD}), as defined by the original authors [3].

$$I_{\epsilon+}(x, x') = \max_{i \in \{1,\dots,n\}} \{f_i(x) - f_i(x')\} \tag{3}$$

$$I_{HD}(x, x') = \begin{cases} H(x') - H(x) & \text{if } x' \succeq x \text{ or } x \succeq x' \\ H(x + x') - H(x) & \text{otherwise} \end{cases} \tag{4}$$

where $x \succeq x'$ indicates x Pareto dominates x'. $I_{\epsilon+}(x, x')$ gives the minimum value by which a solution $x \in \mathcal{P}_t$ has to, or can be translated in the objective space in order to weakly dominate another solution $x' \in \mathcal{P}_t$. $H(x)$ give the multidimensional volume of the objective space that is dominated by x. $I_{HD}(x, x')$ gives the hypervolume that is dominated by x' but not by x, $x, x' \in \mathcal{P}_t$. More information about IBEA can be found in [3].

2.3   The AεSεH

Adaptive ε-Sampling and ε-Hood (AεSεH) [8] is an elitist evolutionary many-objective algorithm that applies ε-dominance principles for survival and parent selection. There is not an explicit fitness assignment method in this algorithm.

Survival selection joins the current population $\mathcal{P}_t$ and its offspring $\mathcal{Q}_t$ and divide it in non-dominated fronts $\mathcal{F} = \{\mathcal{F}_i\}, i = 1, 2, \cdots, N_F$ using the non-dominated sorting procedure. In the rare case where the number of non-dominated solutions is smaller than the population size $|\mathcal{F}_1| < |P|$, the sets of solutions $\mathcal{F}_i$

are copied iteratively to $\mathcal{P}_{t+1}$ until it is filled; if set $\mathcal{F}_i$, $i > 1$, overfills $\mathcal{P}_{t+1}$, the required number of solutions are chosen randomly from it. On the other hand, in the common case where $|\mathcal{F}_1| > |P|$, it calls ε-sampling with parameter ε_s. This procedure iteratively samples randomly a solution from the set $\mathcal{F}_1$, inserting the sample in $\mathcal{P}_{t+1}$ and eliminating from $\mathcal{F}_1$ solutions ε-dominated by the sample. After sampling, if $\mathcal{P}_{t+1}$ is overfilled solutions are randomly eliminated from it. Otherwise, if there is still room in $\mathcal{P}_{t+1}$, the required number of solutions are randomly chosen from the initially ε-dominated solutions and added to $\mathcal{P}_{t+1}$.

For parent selection the algorithm first uses the procedure ε-hood creation to cluster solutions in objective space. This procedure randomly selects an individual from the surviving population and applies ε-dominance with parameter ε_h. A neighborhood is formed by the selected solution and its ε_h-dominated solutions. Neighborhood creation is repeated until all solutions in the surviving population have been assigned to a neighborhood. Parent selection is implemented by the procedure ε-hood mating, which sees neighborhoods as elements of a list that are visited one at the time in a round-robin schedule. The first two parents are selected randomly from the first neighborhood in the list, the next two parents are selected randomly from the second neighborhood, and so on. When the end of the list is reached, parent selection continues with the first neighborhood in the list. Thus, all individuals have the same probability of being selected within a specified neighborhood, but due to the round-robin schedule individuals belonging to neighborhoods with fewer members have more reproduction opportunities that those belonging to neighborhoods with more members.

Both epsilon parameters ε_s and ε_h used in survival selection and parent selection, respectively, are dynamically adapted during the run of the algorithm. Further details about AεSεH can be found in [8].

3   Test Problems, Performance Measures, and Algorithms Parameters

To evaluate the algorithms we use small and large MNK-landscapes [10] randomly generated with $M = 3, 4, 5, 6$ objectives. The small landscapes are defined with $N = 20$ bits and $K = 1$ epistatic bit (5 %). We enumerate these landscapes and analyze the dynamics of the algorithms respect to the optimum set. The size of the Pareto optimal set (POS) found by enumeration and the number of non-dominated fronts are shown in Table 1 under columns $|POS|$ and $Fronts$, respectively. The same table also shows the corresponding fraction (%) of the population sizes $|P|$ to the $|POS|$ for various population sizes investigated. Also, we define large landscapes with $N = 100$ bits and $K = 5$ epistatic bits (5 %) and use them to study the scalability of the algorithms to larger search spaces.

We run the algorithms for a fixed number of T generations, collecting in separate files the sets of non-dominated solutions $\mathcal{F}_1(t)$ found at each generation. The approximation of the POS for a run of the algorithm, denoted $\mathcal{A}(T)$, is built by computing the non-dominated set from all generational non-dominated sets $\mathcal{F}_1(t)$, $t = 0, 1, \cdots, T$, making sure no duplicate solutions are included. In general, the approximation at generation t is given by

Table 1. Size of the Pareto optimal set $|POS|$ and number of *Fronts* in the landscapes with $M = 3, 4, 5$, and 6 objectives, $N = 20$ bits, and $K = 1$ epistatic bit. Fractions $|P|$ / $|POS|$ of population size to the size of the POS (in %) investigated in this study.

M	$|POS|$	*Fronts*	$|P|$ / $|POS|$ (%)		
			50	100	200
3	152	258	32.9	65.8	132.6
4	1,554	76	3.2	6.4	12.9
5	6,265	29	0.8	1.6	3.2
6	16,845	22	0.3	0.6	1.2

$$\mathcal{X}(t) = \{\mathcal{A}(t-1) \cup \mathcal{F}_1(t) \setminus \mathcal{A}(t-1) \cap \mathcal{F}_1(t)\} \tag{5}$$

$$\mathcal{A}(t) = \{\boldsymbol{x} : \boldsymbol{x} \in \mathcal{X}(t) \wedge \, \nexists \boldsymbol{y} \in \mathcal{X}(t) \; \boldsymbol{y} \succeq \boldsymbol{x}\} \tag{6}$$

$$\mathcal{A}(0) = \mathcal{F}_1(0), \tag{7}$$

where $\boldsymbol{y} \succeq \boldsymbol{x}$ denotes solution $\boldsymbol{y}$ Pareto dominates solution $\boldsymbol{x}$.

For small landscapes we report the basic resolution index α of the approximation at generation t [11], expressed by

$$\alpha(t) = \frac{|\{\boldsymbol{x} : \boldsymbol{x} \in \mathcal{A}(t) \wedge \boldsymbol{x} \in POS\}|}{|POS|}, \tag{8}$$

which gives the fraction of the accumulated number of Pareto optimal (PO) solutions found until generation t to the size of the POS. The highest resolution of the generated approximation of the POS is achieved when all Pareto optimal solutions are found. We also report three generational search assessment indices [11], the fraction τ_t^+ of Pareto optimal solutions in the population at generation t that are new respect to the previous generation, the fraction δ_t of Pareto optimal solutions dropped at generation t, and the fraction γ_t of non-dominated solutions in the population that are not Pareto optimal solutions at generation t. Table 2 summarizes these indices.

For landscapes with $N = 100$ bits, where the Pareto optimal set is unknown, we compute the non-dominated reference set $\mathcal{R}$ from the solutions found by all algorithms. We report the Inverse Generational Distance (IGD) between the approximation $\mathcal{A}(T)$ found by the algorithms and the reference set $\mathcal{R}$. In addition, we also report the coverage C metric between the approximations $\mathcal{A}(T)$ found by the algorithms.

All algorithms use two point crossover with rate $pc = 1.0$, and bit flip mutation with rate $pm = 1/N$. In MOEA/D we use the Tchebycheff scalarizing function, as mentioned above, set the neighborhood size to 10, as suggested for knapsack problems in the original implementation of MOEA/D. The set of weights vectors is generated according to the methodology presented in [12], which projects the discrepancy given by a set of points contained in a $(k - 1)$-dimensional unit cube into a $(k - 1)$-simplex that defines the set of weights vectors. One advantage of using this strategy is that we can define a well-distributed set of weights vectors (in terms of low discrepancy) without depending of any constant as conventional methodologies do (see e.g. [2]) and regardless

Table 2. Generational search-assessment indices I_t. Measures are taken on non-dominated population $\mathcal{F}_1(t)$ with respect to $\mathcal{F}_1(t-1)$ and/or the POS, normalized by population size $|P|$.

I_t	Formula	Comment
τ_t^+	$|\{\boldsymbol{x} : \boldsymbol{x} \in \mathcal{F}_1(t) \wedge \boldsymbol{x} \notin \mathcal{F}_1(t-1) \wedge \boldsymbol{x} \in POS\}| \, / \, |P|$	Possibly new PO solutions
δ_t	$|\{\boldsymbol{x} : \boldsymbol{x} \in \mathcal{F}_1(t-1) \wedge \boldsymbol{x} \notin \mathcal{F}_1(t) \wedge \boldsymbol{x} \in POS\}| \, / \, |P|$	Dropped PO solutions
γ_t	$|\{\boldsymbol{x} : \boldsymbol{x} \in \mathcal{F}_1(t) \wedge \boldsymbol{x} \notin POS\}| \, / \, |P|$	Non-dominated, not PO sol

of the dimension of the weights vectors. In AεSεH we set the reference neighborhood size H_{size}^{Ref} to 20 individuals. The mapping function $\boldsymbol{f}(\boldsymbol{x}) \mapsto^\epsilon \boldsymbol{f}'(\boldsymbol{x})$ used for ε-dominance in ε-sampling truncation and ε-hood creation is additive, $f_i' = f_i + \varepsilon, i = 1, 2, \cdots, m$. For IBEA, we observe the behavior of the algorithm setting the scaling factor to $\kappa = 0.05$ suggested in [3] and $\kappa = 0.001$. IBEA finds considerably fewer optimal solutions if $\kappa = 0.05$. Here we report results obtained setting $\kappa = 0.001$. The algorithms run for $T = 100$ generations with population sizes $|P| = \{50, 100, 200\}$ on landscapes with $N = 20$ bits and for $T = 1000$ generation with population size $|P| = 1000$ on landscapes with $N = 100$ bits. Results analyzed here were obtained from 30 independent runs of the algorithms.

4   Experimental Results and Discussion

4.1   Small Landscapes

First, we analyze the basic resolution index $\alpha(T)$ of the approximation at the end of the run, i.e. the ratio of accumulated number of PO solutions found to the size of the POS. Results for all algorithms are shown in Fig. 1 for 3, 4, 5, and 6 objectives using population sizes of $\{50, 100, 200\}$. For convenience the algorithms are labeled as A, Ie, Ihv, M, and N and correspond to AεSεH, IBEA $I_{\varepsilon+}$, IBEA I_{HD}, MOEA/D, and NSGA-II, respectively.

For $M = 3$ objectives, note that AεSεH finds more Pareto optimal solutions than the other algorithms for the three population sizes tried here. MOEA/D finds more Pareto optimal solutions than NSGA-II for population size 50, but the contrary is true for population sizes 100 and 200. IBEA $I_{\varepsilon+}$ and I_{HD} find consistently fewer Pareto optimal solutions than the other algorithms. In $M = 3$ the ratios of population size to the size of the Pareto optimal set are $|P|/|POS| \sim \{33, 66, 133\}$ (%) for $|P| = \{50, 100, 200\}$, respectively. That is, the population size is relatively large compared to the Pareto optimal set. In this case, note that the difference in the resolution achieved by the algorithms reduces considerably as the ratio $|P|/|POS|$ increases to very large values.

On the other hand, for 4, 5 and 6 objectives, note that overall MOEA/D finds more Pareto optimal solutions than the other algorithms, followed by AεSεH. NSGA-II scales up badly in the number of objectives and becomes similar or worse than IBEA $I_{\varepsilon+}$ and IBEA I_{HD}. In $M = 4$ the ratios are $|P|/|POS| \sim$

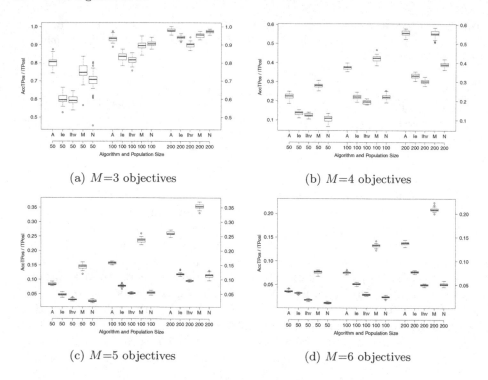

(a) M=3 objectives

(b) M=4 objectives

(c) M=5 objectives

(d) M=6 objectives

Fig. 1. Resolution of the approximation at the end of the run $\alpha(T)$, i.e. ratio of accumulated number of Pareto optimal solutions found to the size of the POS. Population sizes 50, 100, and 200 for 3, 4, 5, and 6 objectives. Algorithms AεSεH (A), IBEA $I_{\varepsilon+}$ (I_ε), IBEA I_{HD} (I_{hv}), NSGA-II (N) and MOEA/D (M).

{3.2, 6.4, 12.9} (%). In this case the advantage of MOEA/D over AεSεH seen for ratios 6.4 % and 3.2 % disappears for the ratio 12.9 % ($|P = 200|$). In $M = 5$ and $M = 6$ the ratios $|P|/|POS|$ used in our experiments are around {0.8, 1.6, 3.2} (%) and {0.3, 0.6, 1.2} (%). These ratios are quite small and the superiority of MOEA/D to achieve a better resolution is undisputed.

In 3, 4, and 5 objectives landscapes with $N = 20$ bits the algorithms can hit easily the Pareto optimal set after few generations. In $M = 6$ there are few optimal solutions even in the random initial population. Therefore, the above results reflect mostly the ability of the algorithms to continue discovering Pareto optimal solutions once they hit the Pareto optimal set.

In the following we analyze the dynamics of the algorithms for $M = 3$ objectives with population size $|P| = 50$, where $|P|$ is 32.9 % of the $|POS|$, and for $M = 6$ with $|P| = 200$, where $|P|$ is 1.2 % of the $|POS|$. Our aim is to understand the behavior of the algorithms under small and large ratios $|P|/|POS|$ and explain how the algorithms achieve the resolutions observed in Fig. 1. This analysis will also help understand how the scalability to larger search spaces could be affected by the dynamics of the algorithms.

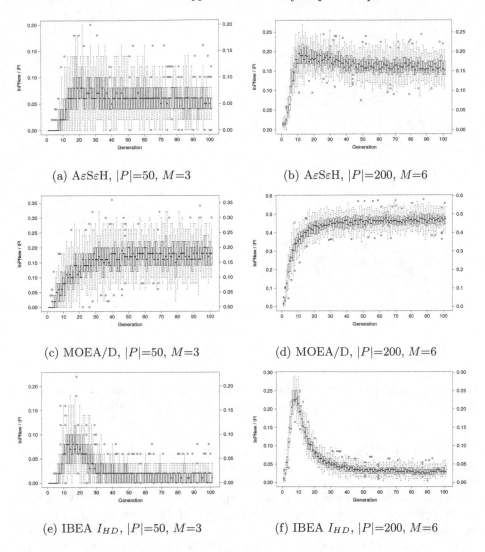

(a) AεSεH, $|P|$=50, M=3

(b) AεSεH, $|P|$=200, M=6

(c) MOEA/D, $|P|$=50, M=3

(d) MOEA/D, $|P|$=200, M=6

(e) IBEA I_{HD}, $|P|$=50, M=3

(f) IBEA I_{HD}, $|P|$=200, M=6

Fig. 2. Pareto optimal solutions in the population that are new respect to the previous generation. Population sizes 50 and 200 for 3 and 6 objectives, respectively. Algorithms AεSεH, MOEA/D, and IBEA I_{HD}.

Figure 2 shows the fraction τ_t^+ of Pareto optimal solutions that are new in the population respect to the previous generation. That is, τ_t^+ includes Pareto optimal solutions that are being rediscovered and also those seen for the first time. Note that τ_t^+ in AεSεH and MOEA/D peak during the initial generations and remain close to its peak value throughout the generations. However, τ_t^+ in AεSεH is smaller than in MOEA/D (around half), both in $M = 3$ with $|P| = 50$ (32.9 % of the $|POS|$) and $M = 6$ with $|P| = 200$ (1.2 % of the $|POS|$).

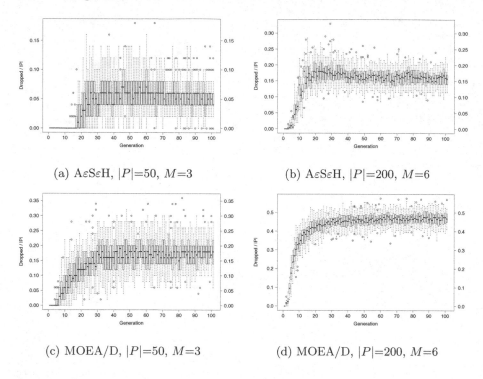

(a) AεSεH, $|P|$=50, M=3

(b) AεSεH, $|P|$=200, M=6

(c) MOEA/D, $|P|$=50, M=3

(d) MOEA/D, $|P|$=200, M=6

Fig. 3. Pareto optimal solutions dropped from the population. Population sizes 50 and 200 for 3 and 6 objectives, respectively. Algorithms AεSεH and MOEA/D.

In the case of IBEA, after τ_t^+ has reached its peak rapidly drops to a very small value, indicating that IBEA rediscovers and/or finds very few new Pareto optimal solutions after 30 generations.

Figure 3 shows the ratio δ_t of Pareto optimal solutions in the population that are dropped over the generations. These dropped solutions are replaced by other non-dominated solutions, optimal or not. Note that the trends of the curves are similar to those of τ_t^+ shown in Fig. 2. MOEA/D drops almost three times as many Pareto optimal solutions as AεSεH in both cases, $M = 3$ with $|P| = 50$ (32.9 % of $|POS|$) and $M = 6$ with $|P| = 200$ (1.2 % of $|POS|$). IBEA drops very few solutions, particularly after the algorithm has evolved few generations (results are not included here due to space limitations).

Figure 4 shows the ratio γ_t of solutions that are non-dominated in the population but are not Pareto optimal. Note that γ_t in AεSεH is larger than in MOEA/D during the initial 20 or 10 generations, where the algorithms are approaching the optimal front and few solutions in the population are expected to be Pareto optimal. However, after this initial period, when a significant number of Pareto optimal solutions should have accumulated in the population γ_t is three times higher in MOEA/D than in AεSεH.

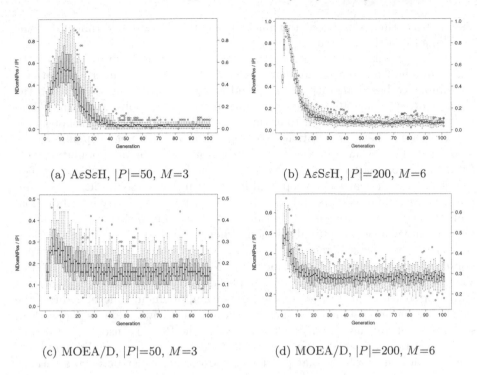

Fig. 4. Non-dominated solutions in the population that are not Pareto optimal. Population sizes 50 and 200 for 3 and 6 objectives. Algorithms AεSεH and MOEA/D.

To summarize, whether the fraction $|P|/|POS|$ is small or large, MOEA/D discovers and rediscovers more Pareto optimal solutions than AεSεH. However, MOEA/D also drops more optimal solutions than AεSεH and includes in its population a larger number of non-dominated non-Pareto optimal solutions than AεSεH. The discovery of new Pareto optimal solutions together with the ability to drop and replace them with other Pareto optimal solutions can be seen as an exploitative feature of the algorithm to continue reaching optimal solutions from optimal solutions. However, Pareto optimal solutions are also replaced with non-optimal solutions. In this case, the algorithm steps down to inferior solutions and tries to climb back again. This feature is more explorative and could help the algorithm to scape local optima, or to reach optimal solutions that cannot be reached easily from other optimal solutions. These two features are observed in both MOEA/D and AεSεH. However, the indices explored here suggest that exploration in MOEA/D is more intense than in AεSεH. The better approximation achieved by AεSεH on 3 objectives, where there are more fronts to be climbed towards the Pareto optimal set, and the better approximations achieved by MOEA/D on larger number of objectives, where there are less fronts to be climbed, are an indication that this explorative feature could impact greatly the performance of the algorithm. In larger search spaces, it is not so simple to hit the Pareto optimal set. There, too much exploration could be detrimental to the performance of the algorithm.

An important question is how the algorithms come to drop Pareto optimal solutions from the population, particularly in favor of inferior solutions. In dominance based algorithms this could happen during truncation when the number of non-dominated solutions obtained from the combined population of parents and offspring is larger than the size of the population. The scope of the Pareto relation between solutions is the population, and not all points in the landscape. Thus, solutions that appear non-dominated in the population may actually be dominated by other solutions in the landscape. For example, when the algorithm hits parts of the optimal front, even if some solutions in the combined population of parents and offspring are optimal others may be suboptimal and still appear non-dominated. In this case, Pareto optimal solutions could be dropped in favor of suboptimal solutions when the subset of surviving solutions is selected, because a dominance based algorithm cannot distinguish between non-dominated solutions. It is important to emphasize that although inferior solutions in the landscape may appear non-dominated by an optimal solution (superior solutions in general) in the population, dominance never reduces the rank of an optimal solution. In general, dominance never reduces the rank of solutions that are superior in the landscape (in the Pareto sense).

In the case of decomposition algorithms, by definition there is a different function for each sub-problem that provides a more strict order between solutions. In a combinatorial problem, the optimal solution for a sub-problem is hopefully a Pareto optimal solution. Other solutions are inferior, even if they are Pareto optimal in the multi-objective landscape. In general, from the Pareto dominance perspective, solutions that are superior in the multi-objective landscape could be ranked lower than inferior solutions. This is an important difference with dominance based approaches. When the algorithm hits the Pareto optimal set, each optimal solution in the population will be associated to one or few subproblems. These Pareto optimal solutions could be dropped in favor of a solution with higher rank in the subproblem, whether this better ranked solution is superior or not in the Pareto sense.

In the case of IBEA, the algorithm tries to introduce a total order between solutions giving higher rank to solutions located towards the ideal point. Thus, IBEA tends to converge towards the subset of solutions with highest rank located in the central region of objective space, which cardinality is the size of the population. Once there, the continuous sampling from that subset could lead to discover other Pareto optimal solutions. However, they will have a rank inferior to those in the population and thus are not eligible to replace optimal solutions. After a while, the algorithm cannot find new solutions from the same set and stagnates. Due to the total order, this algorithm includes features that can help convergence in larger subspaces, thought diversity could still be an issue.

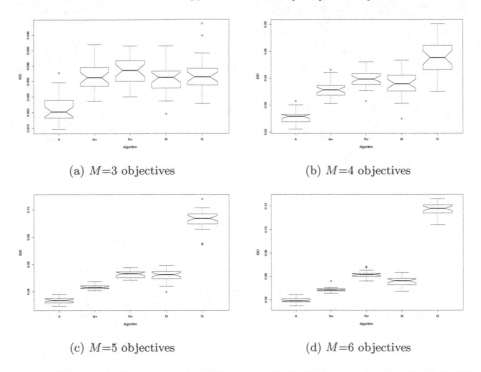

(a) $M=3$ objectives

(b) $M=4$ objectives

(c) $M=5$ objectives

(d) $M=6$ objectives

Fig. 5. IGD. Algorithms AεSεH (A), IBEA $I_{\varepsilon+}$ ($I_{\varepsilon+}$), IBEA I_{HD} (I_{HD}), NSGA-II (N) and MOEA/D (M).

4.2   Large Landscapes

In this section we present results of the algorithms on landscapes with $N = 100$ bits in order to analyze their scalability to larger search spaces. Figures 5 and 6 show the inverse generational distance IGD of the approximation obtained by the algorithms and the coverage C metric between the approximations of AεSεH and the other algorithms, respectively. For these problems we don't know the Pareto optimal set, so we compute IGD taking as reference the non-dominated set obtained from the non-dominated solutions found by all algorithms.

First, looking at IGD in Fig. 5, note that AεSεH achieves better (lower) IGD than the other algorithms in 3, 4, 5 and 6 objectives. In 3 objectives, IBEA $I_{\varepsilon+}$, IBEA I_{HD}, MOEA/D and NSGA-II achieve similar IGD. However, for $M > 3$ objectives IBEA $I_{\varepsilon+}$ is the second best algorithm in terms of IGD. For $M = 4$ and $M = 5$ there is not much difference between IBEA I_{HD} and MOEA/D. However, for $M = 6$ MOEA/D is significantly better than IBEA I_{HD}. NSGA-II is overall the worst algorithm.

Next, looking at coverage C in Fig. 6, note that for $M = 3$ $C(A,\cdot) > C(\cdot,A)$ for all algorithms Ie+, Ihv, M, and N. This indicates that solutions of AεSεH dominate more solutions of the other algorithms and fewer solutions of AεSεH are dominated by solutions of the other algorithms. Increasing the number of

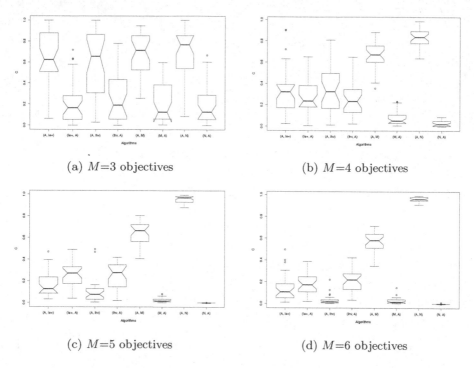

(a) M=3 objectives

(b) M=4 objectives

(c) M=5 objectives

(d) M=6 objectives

Fig. 6. C metric. Algorithms AεSεH (A), IBEA $I_{\varepsilon+}$ ($I_{\varepsilon+}$), IBEA I_{HD} (I_{HD}), NSGA-II (N) and MOEA/D (M).

objectives above 3, the dominance gap between AεSεH and MOEA/D and between AεSεH and NSGA-II increase. However, fewer solutions by IBEA algorithms are dominated by AεSεH. For example in $M = 6$ objectives, in average around 3 % of IBEA I_{HD}'s solutions are dominated by AεSεH and around 20 % of AεSεH's solutions are dominated by IBEA I_{HD}. Between the two IBEA algorithms, C is slightly better for I_{HD} than for $I_{\varepsilon+}$. This however depends strongly on the value set for κ in IBEA.

These results complement our analysis of the previous section and confirms our expectation that too much exploration by MOEA/D could be detrimental to its performance in larger landscapes. It also shows that IBEA can find a subset of well converged solutions. However, it does it at the expense of not finding a well spread set of solutions. AεSεH seems to have a good balance between convergence and diversity, which favors its scalability to larger landscapes. It will be interesting to find ways to control the exploration/exploitation features of the algorithms studied here to improve their performance, whether we scale up the objective space or the search space.

5   Conclusions

This work analyzed and compared the performance of MOEA/D, IBEA using the binary additive ε and the hypervolume difference indicators, and AεSεH for many-objective optimization. We traced the dynamics of the algorithms in small MNK-landscapes, performed and off-line analysis of the Pareto optimal solutions discovered and dropped at each generation, and compared the algorithms for their ability to generate high-resolution approximations of the Pareto optimal set. Our analysis in small landscapes showed that exploration in MOEA/D is more intense than in AεSεH. This favors MOEA/D in small landscapes as we increase the number of objectives, where is relatively easy to hit the Pareto optimal set and exploration is more important to increase the resolution of the approximation. However, in large landscapes too much exploration hinders MOEA/D and AεSεH generates approximations with better convergence and diversity, regardless of the number of objectives. IBEA converges to the central region of objective space, achieving low resolutions in small landscapes. In large landscapes this results in a subset of solutions with very good convergence properties, but poorly spread. In the future we would like to find ways to control the exploration/exploitation features of the algorithms to improve their performance when we scale up the objective and search spaces.

References

1. Hughes, E.: MSOPS-II: a general-purpose many-objective optimiser. In: Proceedings of the IEEE Congress on Evolutionary Computation (CEC), pp. 3944–3951 (2007)
2. Zhang, Q., Li, H.: MOEA/D: a multi-objective evolutionary algorithm based on decomposition. IEEE Trans. Evol. Comput. 11(6), 712–731 (2007)
3. Zitzler, E., Künzli, S.: Indicator-based selection in multiobjective search. In: Yao, X., Burke, E.K., Lozano, J.A., Smith, J., Merelo-Guervós, J.J., Bullinaria, J.A., Rowe, J.E., Tiňo, P., Kabán, A., Schwefel, H.-P. (eds.) PPSN 2004. LNCS, vol. 3242, pp. 832–842. Springer, Heidelberg (2004)
4. Beume, N., Naujoks, B., Emmerich, M.: SMS-EMOA: multiobjective selection based on dominated hypervolume. Eur. J. Oper. Res. 181(3), 1653–1669 (2007)
5. Igel, C., Hansen, N., Roth, S.: Covariance matrix adaptation for multi-objective optimization. Evol. Comput. 15(1), 1–28 (2007)
6. Laumanns, M., Thiele, L., Deb, K., Zitzler, E.: Combining convergence and diversity in evolutionary multi-objective optimization. Evol. Comput. 10(3), 263–282 (2002)
7. Hadka, D., Reed, P.: Borg: an auto-adaptive many-objective evolutionary computing framework. Evol. Comput. 2(2), 231–259 (2013)
8. Aguirre, H., Oyama, A., Tanaka, K.: Adaptive ε-sampling and ε-hood for evolutionary many-objective optimization. In: Purshouse, R.C., Fleming, P.J., Fonseca, C.M., Greco, S., Shaw, J. (eds.) EMO 2013. LNCS, vol. 7811, pp. 322–336. Springer, Heidelberg (2013)
9. Deb, K., Agrawal, S., Pratap, A., Meyarivan, T.: A fast elitist non-dominated sorting genetic algorithm for multi-objective optimization: NSGA-II. KanGAL report, 200001 (2000)

10. Aguirre, H., Tanaka, K.: Insights on properties of multi-objective MNK-landscapes. In: Proceedings of the IEEE Congress on Evolutionary Computation, pp. 196–203. IEEE Service Center (2004)
11. Aguirre, H., Liefooghe, A., Verel, S., Tanaka, K.: An analysis on selection for high-resolution approximations in many-objective optimization. In: Bartz-Beielstein, T., Branke, J., Filipič, B., Smith, J. (eds.) PPSN 2014. LNCS, vol. 8672, pp. 487–497. Springer, Heidelberg (2014)
12. Martínez, S.Z., Aguirre, H., Tanaka, K., Coello, C.: On the low-dyscrepancy sequences and their use in MOEA/D for high dimensionality objective spaces. In: Proceedings of the IEEE Congress on Evolutionary Computation, IEEE Press (2015) (to appear)

Traffic Signal Optimization: Minimizing Travel Time and Fuel Consumption

Rolando Armas$^{(\boxtimes)}$, Hernán Aguirre, Saúl Zapotecas-Martínez,
and Kiyoshi Tanaka

Faculty of Engineering, Shinshu University, 4-17-1 Wakasato, Nagano 380-8553, Japan
rolandoarmas@gmail.com, {ahernan,zapotecas,ktanaka}@shinshu-u.ac.jp

Abstract. This work integrates a multi-objective evolutionary algorithm with the multi-agent transport simulator MATSim and the comprehensive modal emission model simulator CMEM to analyze the evolutionary optimization of traffic signals minimizing travel time and fuel consumption on a real-world large scenario. We simulate the movement of 20.000 vehicles on the transport network of a 5×8 Km^2 area of Quito including 70 signal lights. Our aim is to clarify the nature and the extent of the conflict between these objectives. We also compare with a single-objective optimization algorithm where only travel time is optimized and evaluate the impact of the signals settings on gas emissions.

1 Introduction

The design of sustainable transport systems has received attention in recent years [1]. Population growth and urbanization trends have increased the demand of road networks causing congestion. This adds substantial costs for transportation and business operations, increases the risk of accidents, and increases gas emissions affecting the environment and population health [2]. Sustainable transport systems consider mobility, societal and economic aspects aiming to improve life in urban centers and reduce the impact on the environment.

Designing a sustainable transport system is a highly complex problem. Developments on simulators are helping to create computational models of real world transport systems and emission models. Experts use these simulators to study the transport and mobility system, gain knowledge of it and try alternative hypothesis and scenarios in search of appropriate solutions from a sustainability standpoint. However, the dimension of the problem and the possible number of alternative solutions is overwhelmingly high. Thus, an expert usually focuses on reduced parts of the system and can analyze only a few alternatives that try to solve the problem partially.

Evolutionary computation provides the means to search and explore several alternatives, allowing the expert to direct evolution and focus its analysis on promising solutions found by artificial evolution. Besides, since a sustainable transport system must consider several criteria related to the mobility, the economy, the society and the environment, multi- and many-objective evolutionary approaches

S. Bonnevay et al. (Eds.): EA 2015, LNCS 9554, pp. 29–43, 2016.
DOI: 10.1007/978-3-319-31471-6_3

seem as an appropriate tool to integrate with transport and emissions simulators to help understand the trade-offs inherent to the sustainability problem.

In the literature, there are some works related to single- and multi-objective optimization that partially deal with sustainable transportation systems. For example, Kim et al. [3] solves a road network design problem (RNDP) using a bi-level optimization approach that reflects the different objectives between planners and network users. The authors focused on the design of a very small network with six links and six nodes optimizing three objectives related to travel time, fuel consumption, and accessibility to network's nodes. Stolfi and Alba [4] implemented an evolutionary and rerouting algorithm that suggests alternative routes to avoid traffic jams, showing that it is possible to reduce travel times, greenhouse emissions, and fuel consumption. This approach uses a single-objective optimization algorithm that combines all the criteria into one aggregation function. The authors used four scenarios in a range between 2.5 and 7 Km^2 with a number of vehicles between 1200 to 1400. Traffic lights were considered in the scenarios but were not subject to optimization.

In this work we integrate the Multi-agent Transport Simulator MATSim [5], the Comprehensive Modal Emission Model (CMEM) simulator [6], and a multi-objective evolutionary algorithm. We simulate the movement of 20.000 vehicles on a transport network that covers a significant part of Quito city and includes 70 signal lights. Our aim is to study and understand in a large real-world scenario the influence of optimal signal settings on travel time and fuel consumption. Particularly, we want to clarify the extent of the conflict between these objectives, if any, when they are optimized simultaneously and how the settings of the signals relate to the trade-offs between them. We also compare with a single-optimization algorithm where only travel time is optimized and evaluate the impact of the signals settings on gas emissions.

2   Method

The three main components of the optimization system considered in this study are the Multi-agent Transport Simulator MATSim [5], the Comprehensive Modal Emission Model (CMEM) simulator [6], and a multi-objective evolutionary algorithm. Figure 1 illustrates their interaction.

MATSim allows micro-simulation of agents moving on a transport system producing detailed information about the routes and movements of the agents. MATSim requires as inputs the initial mobility plans for a set of agents and a model of the transport infrastructure. MATSim computes initial routes for the agents based on heuristics and simulates the traffic following the initial plans of the agents. Then it iterates to optimize plans and routes for all agents in order to provide a system in an equilibrium state [7], where no traveler can improve his travel time or utility function by unilaterally changing routes. MATSim can be run with and without traffic lights. If traffic lights are specified, MATSim simulates them microscopically using fixed-time controls [8].

CMEM is a microscopic emissions simulator that computes second-by-second tailpipe emissions and fuel consumption based on different vehicle operating

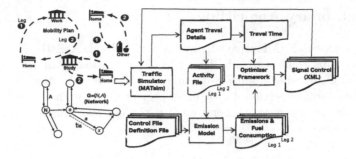

Fig. 1. Optimization system

modes (modal), such as idle, steady-state, cruise, and various levels of acceleration/deceleration [9]. It is called comprehensive because it can predict emissions for a wide range of vehicle / technology categories and various operating conditions, such as properly functioning, deteriorated, malfunctioning. CMEM requires two groups of inputs, input operating variables and model parameters. The input operating variables include information about the activity of the vehicles, that is second-by-second speed (from which acceleration can be derived) and the slope of the road. The model parameters are determined for each one of the vehicles used in the simulation according to the categorization established by CMEM.

Before we run the optimizer, we prepare the initial mobility plans of the agents as well as the model of the transport infrastructure and run MATSim without signal lights until it reaches an equilibrium state. Also, we prepare the profiles of the vehicles associated with the agents, which are required by CMEM.

The multi-objective evolutionary algorithm evolves a population of candidate solutions. Each solution represents the configuration of all light signals (signal control) of the transportation system under study. The algorithm minimizes simultaneously two fitness functions, the average travel time and the fuel consumption of the agents that move in the transport network. At each generation, to compute the fitness of a solution, the evolutionary algorithm calls MATSim and CMEM, one after the other. MATSim sets the signals of the transport system with the values specified by the tentative solution provided by the evolutionary algorithm. Then, MATSim runs one iteration to simulate the movement of the agents following the mobility plans and routes that led the system to the equilibrium state. The output generated by MATSim is used to compute the average travel time of the agents. CMEM is called with the travel details of each agent extracted from the MATSim output and the profiles of the vehicles prepared in advance. The output generated by CMEM is used to compute the fuel consumption of the agents. Once all solutions are evaluated, the evolutionary algorithm continues to the next generation, stopping after a specified maximum number of generations has been completed.

3　Evolutionary Algorithm

In this work we use the Adaptive ε-Sampling and ε-Hood (AεSεH) [10] algorithm to search optimal solutions. AεSεH is an elitist evolutionary multi- and many-objective optimizer that applies ε-dominance [11] principles both for survival selection and parent selection. In the following, we describe the main features of the algorithm, representation, operators of variation, and fitness functions used to study our system.

3.1　AεSεH

AεSεH follows the main steps of a population-based evolutionary algorithm, i.e. parent selection, offspring creation and survival selection, adjusting its operation depending on whether the population contains dominated solutions or not.

　　To perform survival selection, the current population and its offspring are combined and divided into non-dominated fronts using the non-dominated sorting procedure. If the number of non-dominated solutions in the first front is smaller than the population size, the sorted fronts of non-dominated solutions are copied one at the time to the next population until it is filled; if the last copied front overfills the population, the required number of solutions are chosen randomly from it to have the exact number specified by the population size. On the other hand, if the number of non-dominated solutions in the first front is larger than the population size, the first front is truncated to the size of the population using the ε-*sampling* procedure. ε-*sampling* randomly chooses solutions from the first front to include them in the surviving population, eliminating from the front those solutions that are ε-dominated by the chosen samples. As a result, solutions in the next population are spaced according to the $f(x) \mapsto^{\varepsilon_s} f'(x)$ mapping function and parameter ε_s used to compute ε-dominance between solutions.

　　For parent selection, the algorithm first uses a procedure called ε-*hood creation* to cluster solutions in objective space and then applies ε-*hood mating* to select parents. When all solutions in the population are non-dominated, ε-*hood creation* selects *randomly* an individual from the population and applies ε-dominance with mapping function $f(x) \mapsto^{\varepsilon_h} f'(x)$ and parameter ε_h. A neighborhood is formed by the selected solution and its ε_h-dominated solutions. Neighborhood creation is repeated until all solutions in the population have been assigned to a neighborhood. ε-*hood mating* sees the neighborhoods as elements of a list and visits them one at the time in a round-robin schedule. The first two parents are selected *randomly* from the first visited neighborhood in the list. The next two parents are selected randomly from the second neighborhood in the list, and so on. When the end of the list is reached, parent selection continues with the first neighborhood in the list. On the other hand, when dominated solutions are present in the population, ε-*hood creation* makes sure that the solution sampled to create the neighborhood is a non-dominated solution and ε-*hood mating* uses *binary tournaments* based on dominance rank to select parents within the neighborhoods. Both epsilon parameters ε_s and ε_h used in survival selection and neighborhood creation, respectively, are dynamically adapted during the run of the algorithm.

This algorithm has been shown to work effectively on continuous and discrete multi- and many-objective optimization problems [10,12,13]. Further details about the algorithm can be found in [10] and [12].

3.2 Representation

The principal components of a traffic signal are cycle length, phase, offset, stage, green and inter-green time. *Cycle length* is the time in seconds required for one complete color sequence of the signal. A *phase* is the set of movements that can take place simultaneously. An *Offset* is the time lapse in seconds between the beginning of a corresponding green phase at an intersection and the beginning of a corresponding green phase at the next intersection. One *stage* is a green and inter-green time sequence (see Fig. 2).

A signal S in junction h is represented by set of **integer** variables expressed by

$$S_h = (C_h, \theta_h, \phi_{h,1}, \cdots, \phi_{h,r}), \tag{1}$$

where C_h is cycle length, θ_h is the offset, and $\phi_{h,1}, \cdots, \phi_{h,r}$ are the green times for the r phases of the signal. Signal S_h represents one *gene*, and a set of signals constitute the *chromosome* of an individual, i.e. a solution with the complete specification of all signals considered in the system. Figure 3 illustrates the representation of a solution to a system with h signals, each one with two phases. The ranges and constraints of these variables are given in Eqs. (2)–(8), where $I_{h,r}$ is the inter-green time at signal h for phase r and P_h is the total number of phases at signal h. Equations (2)–(4) represent the range for cycle length Ch, offset θ_h and green time $\phi_{h,r}$, respectively. C_{hmin} is determined by identifying

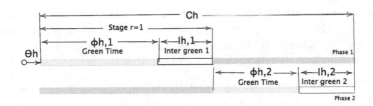

Fig. 2. Traffic light components

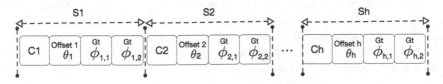

Fig. 3. Chromosome representation

the signal that needs the longest duration just to accommodate the inter-green times and the minimum green times as shown in Eq. (5). C_{max} is set to 135 s. Inter-green per phase is 3 s and minimum green time duration is 17 s for all signals as shown in Eq. (6). These values imply that the minimum cycle time C_{min} is 40 s in two phase signals. Equation (7) ensures that the sum of the green times in a signal together with inter-green do not exceed the cycle length set for the signal. Equation (8) establishes the maximum green time for the signal phase based on the cycle time, inter-green and minimum green time.

$$C_{hmin} \quad \leq \quad C_h \quad \leq \quad C_{hmax} \tag{2}$$

$$0 \leq \theta_h \leq C_h - 1 \tag{3}$$

$$\phi_{h,rmin} \leq \phi_{h,r} \leq \phi_{h,rmax} \tag{4}$$

$$C_{min} = Max \left\{ \left(\sum_{r=1}^{P_h} \phi_{h,r} + \sum_{r=1}^{P_h} I_{h,r} \right) : \quad h = 1, 2..., N \right\} \tag{5}$$

$$\phi_{h,rmin} = 17 \quad sec \quad \forall h, r \tag{6}$$

$$C_h = \sum_{r=1}^{P_h} \phi_{h,r} + \sum_{r=1}^{P_h} I_{h,r} \quad \forall h \tag{7}$$

$$\phi_{h,rmax} = C_h - \sum_{r=1}^{P_h} I_{h,r} - \sum_{y=1, y \neq r}^{P_h} \phi_{h,ymin} \tag{8}$$

3.3   Operators

To create offspring we follow the representation described above and apply crossover with probability P_c and mutation with probability P_m per signal. If a signal undergoes mutation, we apply one of the three mutation operators for cycle length, offset, and green times with probability $P_m^{(Ct)}$, $P_m^{(Of)}$ and $P_m^{(Gt)}$, respectively. The operators are as follows.

Crossover: In this work we implement one point crossover taking each signal as an atomic unit. The crossing point is selected randomly with equal probability in the range $[1, h - 1]$, where h is the number of signals. Then the crossover operator interchanges complete signals between parents.

Cycle Length Mutator: This operator increases or decreases randomly with equal probability the cycle length of a signal using step size $stepCt$. If the new cycle length is out of the specified range, we adjust it accordingly to be either C_{hmin} or C_{hmax}. After that, it is necessary to check whether offset time violates its constraint. If offset is larger than the new cycle length, it is reset to new cycle length $-$ $stepOff$, where $stepOff$ is the offset step size. Finally, for each signal phase the green times are adjusted proportionally to the new cycle length. Due to the correlation of offset and green times to the cycle length, this operator may act as a macro-mutation operator.

Green Time Mutator: This operator decreases the green time of one phase and adds it to another phase using step size $stepGt$. To determine the phase that will decrease its green time, we randomly visit the phases until we find one in which the decrement does not violate the constraint for minimum green time $\phi_{h,rmin}$. The phase to which the green time is added is also determined randomly among all phases, except the one in which time was reduced.

Offset Time Mutator: This operator increases or decreases randomly with equal probability the offset time of a signal using step size $stepOff$. If offset becomes negative, it is reset to 0. Likewise, if offset is greater than the maximum cycle length C_{hmax}, it is reset to $C_{hmax} - stepOff$.

3.4   Fitness Functions

In this work, we minimize two fitness functions, the average travel time and the total fuel consumption of the agents that move in the network. To compute the fitness of a solution, MATSim sets the signals of the system with the values specified by the solution passed by the evolutionary algorithm, simulates the movement of the agents following the routes that led the system to an equilibrium state, and outputs the time taken by each agent to travel each one of the links included in its route. A transport network can be represented by a directed graph $G = (N, A)$, where N represents nodes and A represents links. The travel time for a given vehicle is

$$t_{ia} = t_{ia}^{x} - t_{ia}^{e} \quad a = 1, ..., A; \quad i = 1, ..., V, \tag{9}$$

where t_{ia} represents the travel time on link a for vehicle i, t_{ia}^{x} denotes the time vehicle i exited link a (see Fig. 1), t_{ia}^{e} denotes the time vehicle i entered link a, V is the number of vehicles being simulated, A is the number of links in network, e is the entry node and x is the exit node [14]. Thus, the average travel time, the first fitness function, is expressed by

$$f_1 = \frac{\sum_{i=1}^{V} \sum_{a=1}^{A} t_{ia}}{V}, \tag{10}$$

subject to signal timing design and feasibility constraints shown in Eqs. (2)–(8) [15].

The second fitness function corresponds to the fuel consumption of the agents along their legs. It is computed from the output generated by CMEM, which is called along with the travel details of the agents produced by MATSim and the profiles of the vehicles. The second function is stated by

$$f_2 = \sum_{i=1}^{V} \sum_{j=1}^{L} c_i^{j} \tag{11}$$

where V is the number of vehicles, L the number of legs, and c_i^{j} is the fuel consumption (in grams/km) of the ith vehicle at the jth leg.

4 Simulation Results and Discussion

4.1 MATSim and CMEM Preliminaries

The geographical area of study is a large and important part of Quito (Ecuador). It includes the business district, eight major universities, several hospitals, large malls, two large parks, and one major soccer stadium, covering approximately 5×8 Km^2 as shown in Fig. 7. In this area, the slopes of the pathways are in the range from -15 to 15 degrees. For this experiment, we take into account all the pathways with free speeds in the range from 30 to 80 Km/h. The network has 8192 links and comes from Geofabrik and OpenStreetMap [16]. We use the Digital Elevation Model (DEM) from SavGIS [17] to compute the slopes.

The number of simulated agents is 20.000. The mobility plan for each agent consists of two main trips or legs: (1) from home to work, study, or others and (2) from work, study, or others to home (see Fig. 1). The plans are designed so that all agents move first from each home location to different points along the area of study. Those points are facility locations such as universities, workplaces and others like malls, and parks. In their second trip, the agents move back home. The distribution of home locations, workplaces, and education facilities for the mobility plan have been chosen taking into account census data and a previous mobility study [18].

The scenario includes 70 signal lights located on the main pathways with flows in south-north-south, and east-west-east directions (see Fig. 7). We run the multi-agent transport simulator MATSim for 200 iterations, making sure it reaches a user equilibrium state without setting any traffic signal. The traffic simulation period is for 24 hours. It takes approximately 10 hours of computation time to run MATSim for this number of iterations. Traffic signals are optimized using the equilibrium state as an initial condition.

CMEM uses a total of 55 static parameters to characterize the vehicle tailpipe emissions for the appropriate vehicle/technology category. CMEM defines 24 Light-Duty Vehicle (LDV) categories based on fuel and emission control technology, accumulated mileage, power to weight ratio, emission certification level, and emitter level category. We have selected 4 categories based on two main features: accumulated mileage and emitter level category based on model year distribution according to transportation census data [19]. Table 1 shows the vehicle categories chosen for our scenario. We assign randomly a category to each agent according to the distribution obtained from the census.

Table 1. CMEM vehicle categorization

LDV Categories			
9	Tier 1>50 K miles high power/weight	26	Ultra-Low Emission Vehicle
24	Tier 1>100 K miles	27	Super Ultra-Low Emission Vehicle

4.2   Evolutionary Algorithm Experimental Setup

We use a fixed population size of 20. The initial population is created deterministically as follows. We prepare 20 cycle lengths in the range [40, 135] seconds in steps of 5. All solutions are set with a different cycle length, but all signals of a solution are set to the same cycle length. The offset times of all signals are set to zero and green times per phase are set to the same value according to the cycle length, i.e. green time = (cycle length - inter-green) /2. That is, all signals are synchronized to start at the same time but are not coordinated to allow the uninterrupted flow of vehicles along contiguous signals in the same pathway.

For the operators, we set crossover rate to $P_c = 1.0$ and mutation rate per signal to $P_m = 4/n$, where n is the number of signals. The mutation rates for cycle length, offset and green time operators are $P_m^{(Ct)} = 0.5, P_m^{(Of)} = 0.3$ and $P_m^{(Gt)} = 0.2$, respectively. The mutation steps are set to $stepCt=5$, $stepOff=10$, $stepGt=3$ for cycle, offset and green time, respectively. These mutation steps reduce considerably the search space.

We conduct 10 runs of the algorithm setting the number of generations to 50, use different random seeds but all runs start with the same initial population. To evaluate one individual, it takes in average 4 min to run MATSim and compute the first fitness function, and 16 min to run CMEM and compute the second fitness function.

4.3   Results

Figure 4 (a) shows the Pareto fronts found by the algorithm at generations $\{0, 5, 10, 15, 25, 35, 50\}$ for one of the runs. Fuel consumption is converted to

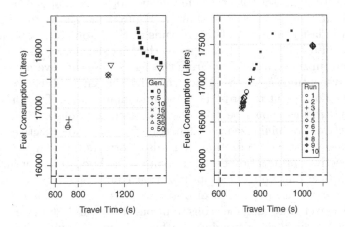

(a) Bi-objective optimization   (b) Single- and bi-objective optimization

Fig. 4. Objective values of solutions by bi- and single-objective optimization

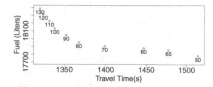

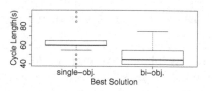

Fig. 5. Non-dominated solutions in initial population labeled with cycle length

Fig. 6. Cycle length, best travel time solutions, single- and bi-objective optimization

liters from kilograms using a gasoline density of 0.755 Kg/liter. The intersection of the dashed lines marks the fitness value of the solution at equilibrium state without signals. Note that a clear trade-off between travel time and fuel consumption can be observed at generation 0 in the initial population. As evolution proceeds, travel time and fuel consumption reduce and approach the values observed at equilibrium state, but the number of non-dominated solutions reduce to a few and in some generations even to one. These results illustrate that the optimization of signals allowing different cycle times and coordinating them by properly setting their offsets lead to significant reductions in both fuel consumption and travel time. The small number of non-dominated solutions in the last generation is expected, because both objective functions are correlated. That is, a reduction in travel time implies that the engines are turned-on for a shorter time and therefore use less fuel.

Here an important question is whether optimizing a single fitness function, either travel time or fuel consumption, could be enough to minimize both objectives. We verify this by optimizing only travel time with an elitist single-objective optimization algorithm [20] set with the same initial population, operators, and parameters used for the bi-objective optimizer. Figure 4 (b) shows the Pareto fronts found at the last generation of the 10 runs of the bi-objective optimization algorithm. It also includes in black squares the best solutions found by the single-objective optimization algorithm. From this figure note that overall results by the bi-objective optimization are better than by the single-objective optimization, thought both point towards the same minimum values. These results suggest that although there could be few non-dominated solutions in the region where both objectives are minimized the inclusion of the second objective helps to perform a more effective optimization. It is also worth noting that variance by the single-objective optimization is larger than by the bi-objective optimization. Nonetheless, the multi-objective optimization could also get trapped in local optima far away from the region of optimality, as observed for run 9 in Fig. 4 (b) where travel time and fuel consumption are around 300 s and 800 liters worse than the solutions with minimum fitness found in run 4. This is a computationally very expensive problem, and not many runs are possible. Thus, it is important to reduce the variance of the solutions found in different runs to increase the reliability of the algorithm. To that end, we should analyze further the operators, population size, and selection of the algorithm in order to find ways to escape local optima.

Figure 4 (a) and (b) illustrate the trade-offs in objective space. In the following, we analyze the settings in decision space, particularly cycle length and offset of the signals. Figure 5 shows the non-dominated solutions in the initial population, fuel consumption over travel time (labeled with cycle length), where all signals of a solution are set to the same cycle length, offset is set to 0, and green times are similar in both traffic flow directions. Note from the figure that when signals are not coordinated, offset set to 0, smaller travel times are achieved by longer cycle lengths and lower fuel consumptions are achieved by shorter cycle lengths. Figure 6 shows box-plots of the cycle length of the best solutions in travel time found by the single- and bi-objective optimization. Note that the optimized solutions include shorter cycle lengths than the best solutions in the initial population and that the cycle lengths by the bi-objective optimization are shorter than by the single-objective optimization. For the single-objective algorithm the highest ranked solution are the ones with the larger cycle length. So, those solutions will be preferred for mating and reproduction. This could imply a loss of diversity of solutions with shorter cycle lengths. However, as indicated above, optimal solutions are a combination of signals with shorter but different cycle lengths. In the case of the bi-objective optimizer, solutions with shorter cycle length will also have a high rank thanks to the second objective, i.e. fuel consumption. Thus, the bi-objective optimizer will not suffer from a lack of diversity of solutions with shorter cycle length. This explains why the bi-objective approach performs a more effective optimization than the single objective approach.

Figure 7 shows the cycle length of the signal lights of the solutions with shorter travel time by the single and bi-objective optimization approaches, deployed on the map of the area of study. Similarly, Fig. 8 shows the offsets of the signal lights. From Fig. 7 it is worth noting that a pattern can be seen in the solutions produced by both approaches. In both solutions, the largest cycle lengths are assigned to signals located in the south-north avenue in the western part of the city. This illustrates the kind of design knowledge we aim to extract from the optimization process, useful to understand and decide the final settings of the signal lights. From Fig. 8 it should be noted that both solutions include some signals with offset 10 or 20, however still many of them remain 0. This is due to the short-term evolution used in this work. The proper setting of offsets undoubtedly helps improve traffic. In the future, we should look for ways to enhance the optimization of offsets.

Table 2 shows travel time and fuel consumption together with HC, CO, NO_x, and CO_2 emissions produced by all agents corresponding to the equilibrium state without traffic signals. Also solutions including traffic signals at generation 0 with smallest values in travel time and fuel consumption, and solutions with traffic signals that minimize travel time by the bi- and single-objective optimizer at generation 50. These results illustrate that in addition to minimizing travel

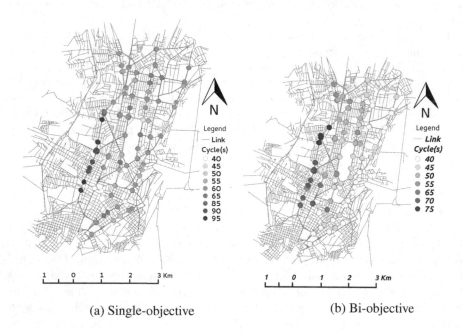

(a) Single-objective   (b) Bi-objective

Fig. 7. Best solution cycle

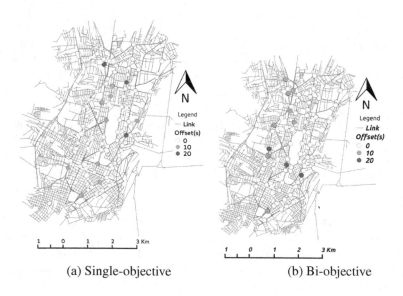

(a) Single-objective   (b) Bi-objective

Fig. 8. Best solution offset

Table 2. Scenario Emissions

	Eq. State	g=0 C_h=130	g=0 C_h=50	g=50 Bi-obj	g=50 Single-obj
Travel Time (s)	608	1320	1513	709	734
Fuel Consum. (l)	15817	18384	17780	16665	16858
HC (Kg)	121.97	129.76	128.26	125.81	126.32
CO (Kg)	2623.90	2764.50	2762.97	2736.81	2737.88
NO_x (Kg)	277.44	253.71	262.34	264.18	261.89
CO_2 (Kg)	33347.37	39248.40	37810.00	35188.40	35647.50

time and fuel consumption, the various kinds of emissions can also be reduced significantly if traffic lights are optimized.

Finally, Fig. 9 shows the traffic volume for the one-day simulation and during peak hours observed for the scenario studied in this work. Note that the main flows of agents go south-north-south rather than east-west-east, which reflects the demographics of the city.

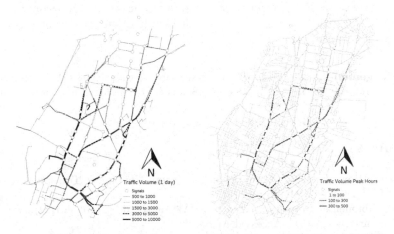

(a) One day simulation.   (b) Peak hours (08h00 and 17h00).

Fig. 9. Traffic volume

5   Conclusion and Future Work

In this work, we analyzed the evolutionary optimization of traffic signals minimizing simultaneously travel time and fuel consumption on a large real-world scenario. We integrated a multi-objective evolutionary algorithm with the transport simulator MATSim and the emissions model simulator CMEM. We used as a case study the transport network of a 5×8 Km^2 area of Quito set with 70 signal lights, and simulated one day traffic of 20.000 agents moving according a two-leg mobility plan. We showed that there is a clear trade-off between travel time and fuel consumption when the signals are set with the same cycle length and are not coordinated (there is not offset between the start of the cycles). We also showed that the optimization of the signals allowing different cycle lengths between signals and coordinating them by properly setting their offsets can reduce significantly both travel time and fuel consumption. This reduces the range of the trade-offs between the two objectives. Further, we verified that the bi-objective optimization approach produces better results than a single-objective approach that optimizes only travel time. We showed evidence that the single-objective algorithm is misled by the initially uncoordinated signals where larger cycle lengths allow shorter travel times, whereas combinations of coordinated signals with shorter cycle lengths lead to better travel times and lower fuel consumption. This was not an issue for the multi-objective optimizers because the second objective related to fuel consumption favors shorter cycle lengths even in uncoordinated signals.

As future works, we should improve the evolutionary algorithm to reduce its variance and enhance its reliability for short-term evolution and few fitness evaluations. Also, we should study other mobility plans and scenarios for the agents. Furthermore, in addition to optimizing traffic signals, we would like to add new variables and optimization criteria to study other important aspects of sustainable transport systems.

Acknowledgements. The first author gratefully acknowledges the support of National Secretariat of Higher Education, Science, Technology and Innovation of Ecuador.

References

1. United Nations, Economic Commission for Europe. Intelligent Transport Systems (ITS) for Sustainable Mobility (2012)
2. US Environmental Protection Agency, EPA. Air trends (2010). http://www.epa.gov/air/airtrends/2010/. Accessed August 2014
3. Kim, J.H., Bae, Y.K., Chung, J.H.: Multi objective optimization for sustainable road network design problem. In: Proceedings of International Conference on Transport, Environment and Civil Engineering (ICTECE), pp. 104–108 (2012)
4. Stolfi, D.H., Alba, E.: Eco-friendly reduction of travel times in european smart cities. In: Proceedings of Conference on Genetic and Evolutionary Computation (GECCO). ACM, pp. 1207–1214 (2014)
5. Multi agent transport simulation (MATSim). http://matsim.org. Accessed January 2014
6. Comprehensive modal emission model (CMEM). http://www.cert.ucr.edu/cmem/index.html. Accessed January 2014
7. Wardrop, J.G.: Some theoretical aspects of road traffic research. ICE Proc. Eng. Divisions **1**(3), 325–362 (1952)

8. Grether, D., Neuman, A.: Traffic light control in multi-agent transport simulations. Technical report, Transport Systems Planning and Transport Telematics, Technical University Berlin (2011)
9. Scora, G., Barth, M.: Comprehensive Modal Emission Model (CMEM), User's Guide version 3.01. University of California Riverside Center for Environmental Research and Technology (2006)
10. Aguirre, H., Oyama, A., Tanaka, K.: Adaptive ε-sampling and ε-hood for evolutionary many-objective optimization. In: Purshouse, R.C., Fleming, P.J., Fonseca, C.M., Greco, S., Shaw, J. (eds.) EMO 2013. LNCS, vol. 7811, pp. 322–336. Springer, Heidelberg (2013)
11. Laumanns, M., Thiele, L., Deb, K., Zitzler, E.: Combining convergence and diversity in evolutionary multiobjective optimization. Evol. Comput. **10**(3), 263–282 (2002)
12. Aguirre, H., Yazawa, Y., Oyama, A., Tanaka, K.: Extending AεSεH from many-objective to multi-objective optimization. In: Dick, G., Browne, W.N., Whigham, P., Zhang, M., Bui, L.T., Ishibuchi, H., Jin, Y., Li, X., Shi, Y., Singh, P., Tan, K.C., Tang, K. (eds.) SEAL 2014. LNCS, vol. 8886, pp. 239–250. Springer, Heidelberg (2014)
13. Aguirre, H., Liefooghe, A., Verel, S., Tanaka, K.: An analysis on selection for high-resolution approximations in many-objective optimization. In: Bartz-Beielstein, T., Branke, J., Filipič, B., Smith, J. (eds.) PPSN 2014. LNCS, vol. 8672, pp. 487–497. Springer, Heidelberg (2014)
14. Spiegelman, C., Sug-Park, E., Rilett, L.: Transportation Statistics and Microsimulation. CRC Press Taylor and Francis Group, Abingdon (2011)
15. Teklu, F., Sumalee, A., Watling, D.: A genetic algorithm approach for optimizing traffic control signals considering routing. Comput. Aided Civil Infrastruct. Eng. **22**(1), 31–43 (2007)
16. Frederik, R., Topf, J., Karch, C.: Geofabrik (2007). http://www.geofabrik.de. Accessed January 2014
17. Souris, M.: Institut de Reserche pour le Developpement (IRD) (2014). http://www.savgis.org/ecuador.htm#DEM30. Accessed October 2014
18. Demoraes, F.: Movilidad, elementos esenciales y riesgos en el Distrito Metropolitano de Quito. PhD thesis, Universidad de Savoie - Francia (2005)
19. National Institute of Statistics of Ecuador (INEC) (2010). http://www.ecuadorencifras.gob.ec/. Accessed October 2014
20. Armas, R., Aguirre, H., Tanaka, K.: Effects of mutation and crossover operators in the optimization of traffic signal parameters. In: Dick, G., Browne, W.N., Whigham, P., Zhang, M., Bui, L.T., Ishibuchi, H., Jin, Y., Li, X., Shi, Y., Singh, P., Tan, K.C., Tang, K. (eds.) SEAL 2014. LNCS, vol. 8886, pp. 167–179. Springer, Heidelberg (2014)

How to Mislead an Evolutionary Algorithm Using Global Sensitivity Analysis

Thomas Chabin$^{(\boxtimes)}$, Alberto Tonda, and Evelyne Lutton

UMR 782 GMPA, INRA, 1 Av. Lucien Brétignères, 78850 Thiverval-Grignon, France
{thomas.chabin,alberto.tonda,evelyne.lutton}@grignon.inra.fr

Abstract. The idea of exploiting Global Sensitivity Analysis (GSA) to make Evolutionary Algorithms more effective seems very attractive: intuitively, a probabilistic analysis can prove useful to a stochastic optimisation technique. GSA, that gathers information about the behaviour of functions receiving some inputs and delivering one or several outputs, is based on computationally-intensive stochastic sampling of a parameter space. Nevertheless, efficiently exploiting information gathered from GSA might not be so straightforward. In this paper, we present three mono- and multi-objective counterexamples to prove how naively combining GSA and EA may mislead an optimisation process.

1 Introduction

Sensitivity analysis is the study of how the uncertainty in the output of a mathematical function can be apportioned to different sources of uncertainty in its inputs [19]. In general, Sensitivity Analysis can be applied to any function f, $\mathbb{R}^n \to \mathbb{R}^p$. In practice, this technique is widely exploited by the modeling community, to analyze the behaviour of models with respect to their parameters, and to later plan new experiments to reduce the uncertainty on the most sensitive parameters. Indeed, a model can be defined as a function $f : X_l, K_n \to Y_m$, whose objective is to simulate a real physical phenomena. Knowing the initial conditions represented by the vector X_l, the model produces the final conditions of the studied phenomena, Y_m. In real-world cases, the parameters of the function K_n are not known with precision but rather defined by a range value of uncertainty. Many sensitivity analysis tools perform a stochastic sampling of considerable magnitude in the space of parameters, and then exploit statistical techniques to derive information from this large quantity of data.

It is easy to see the potential interest of data collected through sensitivity analysis for an optimisation of the parameters of the model: not only sensitivity analysis provides a fine-grained sampling of a search space, but it also conveys useful information about how each parameter influences each output. This holds true especially for evolutionary optimisation techniques, that are

This work has been funded by the French National Agency for research (ANR), under the grant ANR-11-EMMA-0017, EASEA-Cloud Emergence project 2011, http://www.agence-nationale-recherche.fr/.

© Springer International Publishing Switzerland 2016
S. Bonnevay et al. (Eds.): EA 2015, LNCS 9554, pp. 44–57, 2016.
DOI: 10.1007/978-3-319-31471-6_4

based on a biased stochastic sampling of the search space. Re-using the extensive amount of computation performed for a sensitivity analysis to improve the performance of an evolutionary algorithm (EA) targeting the same search space, sounds not only sensible, but also extremely appealing. Not surprisingly, the literature already shows approaches that exploit the synergy between sensitivity analysis and EAs [7]. However, making use of the information conveyed by sensitivity analysis might not be as straightforward as it seems.

In this paper, we exhibit three case studies, specifically designed to deceive an EA exploiting sensitivity analysis data. Experimental results show that even a state-of-the-art EA is unable to find the optimal parameter configuration for the problems, if biased by the information provided by sensitivity analysis; on the contrary, the same algorithm routinely converges on the global optimum if no aprioristic knowledge is given, thus proving that a naive use of sensitivity analysis information might actually be harmful to the optimisation process.

The rest of the paper is organized as follows: Sect. 2 recalls a few basic concepts of sensitivity analysis, with a particular focus on the analysis of joint variation of parameter interactions, and lists previous works at the interface of sensitivity analysis and EAs. Section 3 discusses one of these combination strategies. Counterexamples and experimental results are illustrated in Sect. 4, while the implications are discussed in Sect. 5. Finally, Sect. 6 concludes the paper.

2 Background

2.1 Sensitivity Analysis: Global and Local

Sensitivity analysis is a technique used to understand how variation in the output of a function can be apportioned qualitatively or quantitatively to different uncertain input sources. Sensitivity analysis techniques can be broadly classified as *local* or *global*. Local sensitivity analysis (LSA) is the simpler approach, where only one function variable is perturbed at a time, while the remaining are fixed to a nominal value. Different studies have shown that limiting the analysis to local sensitivities might deliver unreliable results [20, 23]. Thus, global sensitivity analysis (GSA) [19] that examines the joint variation of variable interactions, seems to be better suited for complex, nonlinear models.

2.2 Global Sensitivity Analysis

GSA is mainly used for two goals: *factor prioritizing*, deciding which variable uncertainty to work on, in order to reduce the uncertainty of the output; and *factor fixing*, highlighting which variables can be fixed to an arbitrary value with few influence on the output. One of the most common approaches has been developed by Sobol [21]. Impacts of each individual decision variable and its interactions with other variables on performance objectives are represented with the following sensitivity indices, taking values in $[0, 1]$.

First-order sensitivity indices are used for the factor priority problem. A first-order index S_i is associated to each parameter K_i, and represents the direct influence of its uncertainty on an output Y:

$$S_i = \frac{V[E(Y|K_i)]}{V(Y)}$$

It corresponds to the part of the variance of Y explained directly by the uncertainty in K_i: $V[E(Y|K_i)]$ is the conditional expectation of Y knowing K_i, fixed at each possible value within the uncertainty range of K_i. Fixing to its true value the variable associated to the highest first-order index, would lead to the greatest reduction in the variance of the output.

Higher-orders sensitivity indices correspond to interaction effects. For instance, indices of order 3 S_{ijk} are associated to each triplet of parameters K_i, K_j, K_k:

$$S_{ijk} = \frac{V[E(Y|K_i, K_j, K_k)]}{V(Y)}$$

The sum of all n-order indices is always equal to 1. The computation of higher-order indices is expensive, as there are $\binom{n}{k}$ of such indices for k parameters. In practice, they are rarely used. They are not considered in this paper.

Total-effect sensitivity indices are used for the factor-fixing problem. A total-effect index is attributed to each parameter, and it is interpreted as the sum of all n-order indices involving the considered parameter. A total effect index S_{Ti} represents how much the uncertainty of a parameter, combined with every other uncertainty, is responsible for the output variance:

$$S_{Ti} = 1 - \frac{V[E(Y|K_{\sim i})]}{V(Y)}$$

$K_{\sim i} = K_1, K_2, ..., K_{i-1}, K_{i+1}, ...K_n$ is the set of all parameters except K_i. Therefore, if a parameter has a total-effect index near zero, its uncertainty has nearly no influence on the output variance. For this reason, this parameter can be fixed to an arbitrary value inside his interval of uncertainty without affecting much the variance of the output.

2.3   Sensitivity Analysis and Optimisation

In order to compute GSA indices, the search space of a group of parameters is sampled, aiming at finding the parameters whose variation influences the output of a function (or a model) the most. It is therefore not surprising that several attempts have been performed to combine Sensitivity analysis with optimisation tools, especially those featuring a stochastic sampling of the search space.

A considerable number of research lines exploit LSA to perform what is termed *robust optimisation* [2], a set of techniques which seek a certain amount of robustness against uncertainty, seen as variability in the value of the parameters of the problem or its solution. Some work, like [1] also propose a multi-objective strategy to assess the identifiability and LSA of the parameters of a system.

In [22], EAs are used to find the worst possible parameter settings for a model, maximising the distance between experimental data and model predictions. The results are then exploited to evaluate the influence of each parameter on the outputs. While surely interesting, this approach lacks the statistical support of Global Sensitivity Analysis, providing the user with a general impression of the most influential parameters.

Another research line, presented in two technical reports [16,17], aims at using the points sampled by a CMA-ES algorithm [11] during the optimisation process as the basis for a sensitivity analysis, through a *de-biasing* of the sampling. In practice, weights are used on the sampling points, on the basis of the covariance matrix' determinant at each generation, to express their bias with respect to a completely random process. This methodology raises several theoretical questions that will need to be thoroughly analyzed before its widespread application.

In [7], the authors present an example where the use of GSA improves the EA efficiency. They use GSA measurements to reduce the problem's dimensionality, first optimising the values of a sub-set of the most sensitive parameters, and then restarting the evolution from the solutions found in this way, finally optimising the remaining values. However, preliminary results presented in [3] hint that this strategy may not always be viable.

3   Adaptive Dimensionality Reduction Based on GSA

The idea of using progressive refinements techniques to perform a search in high dimensional spaces appeared as attractive for a long time. This very simple idea is at the origin, for instance, of the *messy genetic algorithm* scheme proposed by Goldberg et al. 25 years ago [8]: "Nature did not start with strings of length two million (an estimate of the number of genes in Homo sapiens) and try to make man. Instead, simple life forms gave way to more complex life forms, with the building blocks learned at earlier times used and reused to good effect along the way." Messy GAs rely on a variable length bit-string representation of the search space made of a list of couples (locus, allele value) specifying the value of a bit at a given place of the genome. In this way some genes may be over-specified (several possible values) while other may be under-specified (no affected value). Fitness calculation is then performed after an additionnal stage relying on various rules for inferring uncomplete string values. This scheme has been extended in various ways including continuous search spaces [12,18]. It implements a self-adaptive progressive refinement, where the selection of primary, "heavy" parameters, is let to evolution.

Adaptive schemes (in the sense of "non-self-adaptive") may also be considered in this context, the critical point being an *a priori* knowledge of an importance prioritization of the parameters. Sensitivity analysis may then represent an attractive solution to deal with parameters importance ordering. The idea is to identify non-influential parameters, via a sensitivity analysis of the fitness function with respect to each parameter in the search space. A straightforward strategy for dimensionality reduction is then to ignore non-influential parameters in a first optimisation stage, like in [7].

4 Experimental Analysis

We propose a series of counterexamples for testing the limits of dimensionality reduction based on GSA, in the same spirit as deceptive functions design [9,10]: global information collected through statistical analysis of some features (building blocks statistics in the case of deceptiveness "à la Goldberg") yields puzzling information to the algorithm. Other interpretations may also stem from theoretical studies regarding the influence of local regularity features [13,15]: global optima are located in very irregular areas, while attractive local optima are located inside smooth areas. Statistical features are actually not able to capture local irregularities and are thus yielding erroneous information to the algorithm [14].

The strategy that is tested relies on the following statement (factor fixing approach, see Sect. 2.2): *a low total effect index indicates a non-influential parameter that can be arbitrarily fixed with only few impact on the fitness function.* To decide which parameters are non-influential, a threshold is arbitrarily fixed (a low value in the range $[0,1]$): parameters that have a total sensitivity index below this threshold are considered non-influential.

4.1 Algorithms

Three EAs have been tested: (i) CMA-ES, (ii) an explicit population based EA, implemented with the EASEA package[1] [4] and (iii) NSGA-II, a multi-objective genetic algorithm. The following schemes have been considered for progressive refinement:

- *Approach 1* performs an optimisation of the influential parameters only. Non-influential parameters are fixed to the middle of their interval of uncertainty.
- *Approach 2* is based on [7]. Influential parameters are optimised in a first stage, like in Approach 1, and then the best point is injected in the initial population of a second optimisation, this time using all parameters.

CMA-ES. The *Covariance Matrix Adaptation Evolution Strategy* (CMA-ES) [11] is a popular EA, widely used for many real-world optimisation problems. It is known for its robustness and computational efficiency. For Approach 2, CMA-ES is restarted as follows:

- The mean point is initialised to the best set of influential parameters found during the first stage, while the values of non-influential parameters are set to the middle of their interval of uncertainty.

---

[1] http://easea.unistra.fr.

– The standard deviation for each influential parameter is kept to the value obtained at the last generation of the first stage, and the standard deviation for non-influential parameters is set to $0.3 \times (range_{max} - range_{min})$.

EA. The second algorithm used in our tests is a classical EA, i.e. an explicit population based EA, programmed in EASEA [4]. For Approach 2, the initial population of the second stage is seeded with the content of the last generation of the first stage. The non-influential parameters who were fixed at the middle of their interval of uncertainty (or search space) are attributed a random value in their range of uncertainty.

NSGA-II. The Nondominated Sorting Genetic Algorithm [6] is a Multiobjective evolutionary algorithm. This algorithm builds a set of non-dominated solutions that approximates an optimal Pareto front. Thanks to a clever ranking and to the use of a crowding distance, the population stabilises on an efficient sampling of the Pareto front. Approach 2 with NSGA-II uses a similar setting as above, for the EASEA-EA.

4.2 Counterexample I

The first counterexample is a function for which a non-influential parameter remains important for the precise location of a global optimum. This can be achieved with functions having simultaneously waves along some axes (corresponding to influential "shapes") and thin peaks along other axes. The projection of the fitness function on the subspace of non-influential parameters then provides an averaged viewpoint on the fitness landscape that conceals high, thin peaks. We thus propose the following bi-dimensional function (Fig. 1):

$$fit_1(k1, k2) = g(k1, 1.33, -0.5, 0.15) + g(k2, 7.98, 0.0005, 0.000025) + h(k1)$$

where g is a Gaussian: $g(k, a, b, c) = a \cdot exp(-\dfrac{(k-b)^2}{2c^2})$ and $k1, k2 \in [-1; 1]$

To make optimisation easier with respect to parameter $k1$, a small gradient, $h(k1)$ is added to the fitness:

$$h(k1) = \begin{cases} \frac{1}{1.0005}k1 + \frac{1}{1.0005} & \text{for } k1 \leq 0.0005 \\ -\frac{1}{0.9995}k1 + \frac{1}{0.9995} & \text{elsewhere} \end{cases}$$

A global sensitivity analysis, whose results are presented in Fig. 1 reads that $k1$ is influential whereas $k2$ is not, since the total effect index of $k2$ is far lower that the total effect index of $k1$.

Approach 1 is tested: optimisation is run on parameter $k1$ only, and the result is compared to an optimisation on parameter $k2$ only. Since $k1$ seems to bear all influence whereas $k2$ appears to be non-influential, it is naively expected that the optimisation on $k1$ will find a better value than the optimisation on $k2$.

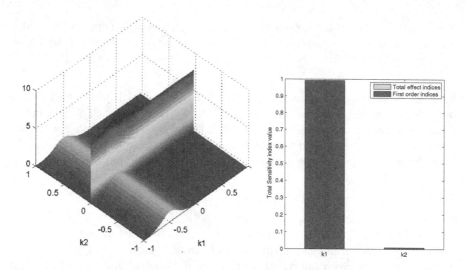

Fig. 1. *Counterexample I.* **(left)** In the fitness landscape, the peak of fit_1 is at $k2 = 0.0005$. The line $k2 = 0$ is at the bottom of the peak. **(right)** Sensitivity analysis shows that $k1$ is much more influential than $k2$.

Table 1. Settings for the EAs used in Counterexample I

	EASEA-EA	CMA-ES
Population size	$\mu = 200$	10
Offsprings size	$\lambda = 180$	-
Number of generations	35	632
Tournament selection	$Size = 2$	-
BLX-α Crossover	$p = 1.$	-
Log normal self adaptive mutation	$p = 1.\tau = \sqrt{2}$	-
Number of Runs	100	100

The algorithms' settings are reported in Table 1. Statistics on 100 runs are displayed in Fig. 2 for the EASEA-EA and CMA-ES algorithms. In this case, optimising on the non-influential parameter is unexpectedly a better option than optimising on the supposedly most influential parameter.

4.3   Counterexample II

A restart strategy (Approach 2 of Sect. 4.1) may counterbalance the problems presented above. We will see however that a restart strategy using GSA may still be puzzled. This is the purpose of counterexample II (Fig. 3).

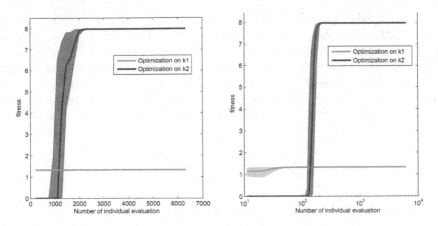

Fig. 2. *Counterexample I.* Comparison of optimisation runs on $k1$ and $k2$, respectively, using the EASEA-EA**(left)** and CMA-ES **(right)**. Statistics on 100 runs are displayed with median in bold and first- and third- quartile in thin lines of the same color.

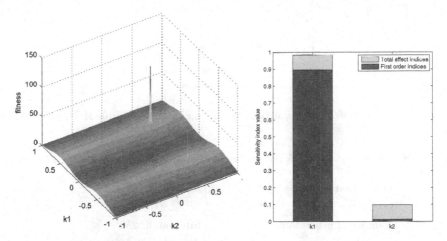

Fig. 3. *Counterexample II.* **(left)** fit_2 has two thin peaks, a very thin one corresponding to a local optimum at $(-0.5, 0.5)$ and a larger one, global optimum, at $(0.5, 0.5)$. **(right)** Sensitivity analysis shows that the total effect index for $k1$ is much higher than for $k2$.

$$fit_2(k1, k2) = g(k1, 10.9, 0.5, 0.25) + g(k1, 11, -0.5, 0.25) + g(k2, 1, 0.5, 0.25)$$
$$+ g2d(k1, k2, 100, 0.5, 0.01, 0.5, 0.01) + g2d(k1, k2, 50, -0.5, 0.0025, 0.5, 0.0025)$$

$k1, k2 \in [-1; 1]$, g and $g2d$ are Gaussians:

$$g(k, a, b, c) = a \cdot exp(-\frac{(k - b)^2}{2c^2})$$
$$g2d(k1, k2, a, b, c, d, e) = a \cdot exp(-(\frac{(k1 - b)^2}{2c^2} + \frac{(k2 - d)^2}{2e^2}))$$

fit_2 has a local optimum at $(k1 = -0.5; k2 = 0.5)$, and a global optimum at $(k1 = 0.5; k2 = 0.5)$. A GSA on Counterexample II (See Fig. 3), shows that $k1$ can be considered as an influential parameter and $k2$ as a non-influential one.

A progressive refinement strategy (Approach 2) is compared to a plain optimisation (full search space) using a classical EA, with the settings reported in Table 2. Over 100 runs, the full search always finds the global optimum whereas the restart strategy (Approach 2) always get stuck on the local optimum (Fig. 4).

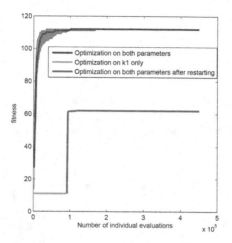

Fig. 4. *Counterexample II.* Statistics of 100 runs on Counterexample II with a classical EA.

Table 2. *Counterexample II.* EA parameter setting, full search space and Approach 2.

Population size	$\mu = 2000$
Offsprings size	$\lambda = 1800$
Number of generations	full search: 250
	Approach 2: 50 then 200
Tournament selection	$Size = 2$
BLX-α Crossover	$p = 1.$
Log normal self adaptive mutation	$p = 1.\tau = \sqrt{2}$
Number of Runs	100

This behaviour is due to the fact that the function is deceptive: when considering only $k1$ for optimisation, and fixing $k2$ to 0, the function has a maximum of 11.14 for $k1 = -0.5$ and a local maximum of 11.04 for $k1 = 0.5$. Thus, the first-stage optimisation concentrates the population around the line $k1 = -0.5$, which prevents the second stage from finding the global peak positioned at $k1 = 0.5$.

The same set of experiments has been performed using CMA-ES with two settings: a first one letting the CMA-ES self-tune its population size, the second one using a larger population size with the idea of artificially maintaining diversity. The results are not reported here, but in both cases, we noticed that Approach 2 was bringing deceiving information to the algorithm, and delayed or even prevented convergence.

4.4   Counterexample III

The third counterexample is based on a multi-objective problem, to better shed light on the potential limits of the method presented in [7]. A bi-objective minimisation problem on a two parameters space has been derived using the fit_2 function. A small offset has been put on parameter $k1$ for the second objective, as follows:

$$fit_{Obj_1}(k1, k2) = -fit_2(k1, k2)$$

$$fit_{Obj_2}(k1, k2) = -fit_2(k1 + 0.05, k2)$$

The theoretical Pareto front is located in the $(k1, k2)$ parameter space, on the segment $[(0.45, 0.5); (0.5, 0.5)]$. A sub-optimal Pareto front also exists on the segment $[(-0.55, 0.5); (0.5, 0.5)]$.

As expected, a GSA on Counterexample III provides similar information on the behaviour of the two objective functions as for Counterexample II: $k1$ is influential on both objectives whereas $k2$ is not (See Fig. 5).

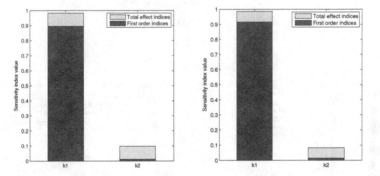

Fig. 5. *Counterexample III.* Sensitivity analysis on objective 1 **(left)** and on objective 2 **(right)**. For both objectives, the total effect index for $k1$ is much higher than for $k2$.

The restart strategy is compared to a classical approach, using the NSGA-II algorithm. The settings for NSGA-II are given in Table 3. The restart strategy always ends up near the sub-optimal Pareto front, whereas the classic strategy finds solutions near the optimal Pareto front. A typical result is displayed in Fig. 6.

For facilitating comparison, two performance metrics have been computed on 100 runs (see Fig. 7). The hypervolume indicator [24] computes the volume of the dominated portion of the objective space. A high hypervolume value means that the solutions are well spread along the objective space and/or are close to the

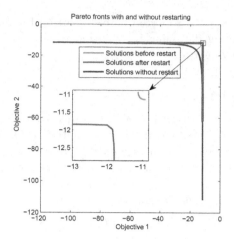

Fig. 6. *Counterexample III.* Typical Pareto front obtained with a classical NSGA-II and with the two steps restart strategy.

Table 3. Settings for NSGA-II on Counterexample III.

Population size	250
Number of generations	full search: 250
	Approach 2: 50 then 200
Number of Runs	100

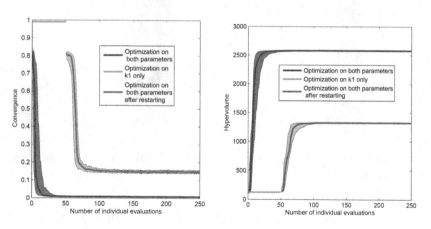

Fig. 7. *Counterexample III.* Convergence metric (**left**) and hypervolume metric (**right**) averaged on 100 runs using NSGA-II and a population size of 250.

optimal Pareto front. The convergence indicator [5] computes a distance between the current solution front and a predefined set of good solutions. Here, solutions have been taken on the theoretical Pareto front. A low value corresponds to a good approximation of the Pareto front.

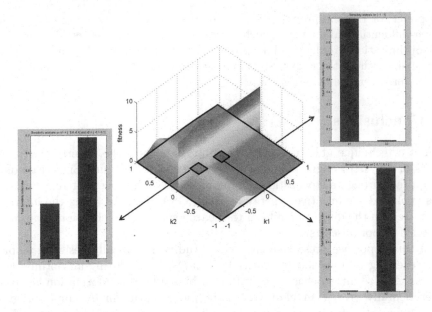

Fig. 8. Various sensitivity analyses on three sub-spaces for Counterexample I: parameters influences vary a lot !

5 Discussion

The counterexamples presented in Sect. 4 shed light on the fact that sensitivity analysis techniques may deliver misleading information to the optimisation process. A possible explanation is that GSA is based on a statistical analysis over a given parameter range. In this way it provides an averaged viewpoint on each parameter, and it is clear that averaging may hide many fine details that are important for optimisation purposes. Another problem is due to the fact that the results of a GSA may drastically vary with the choice of the parameter range. It often happens that a parameter is influential on some subspace and not on another. Figure 8 illustrates this effect for Counterexample I: when $k1, k2 \in \left[-1; 1 \right]$, $k1$ is the parameter that has almost all the influence, whereas $k2$ is almost non-influential. But on other areas, results can be the opposite: for instance if $k1, k2 \in \left[-0.1; 0.1 \right]$, $k1$ is regarded as non-influential, while $k2$ becomes predominant.

The question of an efficient use of GSA inside an optimisation procedure is raised: GSA is, in itself, extremely time consuming, and this cost has not been taken into account in the previous experiments. It seems obvious that GSA, based on a stochastic sampling of the full search space or of an area of it, consumes a computational time that may sometimes be better spent by performing an optimisation process. Additionally, the averaged information provided by GSA may hide some interesting irregular areas where global optima could be found. Finally, adaptive refinement methods, like Approach 2 presented in this paper, or the one proposed in [7], need to identify a non-negligible subset of non-influential parameters, which is not always the case, especially for complex optimisation problems. More progressive strategies may be imagined, but once again with all the risks tied to an assessment of the relative importance of parameters averaged over a given area.

6   Conclusions

GSA is a technique able to deliver information on how the uncertainty in the inputs of a system might influence uncertainty in its outputs. Since this data is acquired through a stochastic sampling of the search space, different research lines exploited the intuitive synergy between GSA and EAs, using the information to reduce the dimensionality of the search space, or to choose the variables on which to optimise first.

In this paper, we presented three case studies, specifically designed to provide deceiving information to sensitivity analysis used during an optimisation process. As a result, stochastic optimisation biased by this information has been experimentally proven unable to reach the global optimum. A simple progressive refinement optimisation scheme based on parameter prioritisation such as in [7] may work on some functions, but there is a risk of falling into a local optimum, from which escaping might prove to be hard. Even if parameter prioritisation might work better for multi-objective problems, thanks to a better diversity preservation mechanism necessary for a correct sampling of Pareto fronts, a multi-objective counterexample is still rather easy to design. This was the purpose of counterexample III.

An interesting point for further developments could be to determine in which cases GSA is beneficial. From this study we can conjecture that regularity of the fitness function may play an important role. If *global* sensitivity analysis has been proven to be puzzling to optimisation in some cases, *local* sensitivity analysis however remains interesting. Sobol indices computed locally for instance may be useful for tuning mutations, in the same spirit as what has been developed in [14], but with an associated computational cost to be taken into account.

References

1. Barichard, V., Hao, J.K.: Resolution d'un probleme d'analyse de sensibilite par un algorithme d'optimisation multiobjectif. In: 5eme Conference Francophone de Modelisation et SIMulation (MOSIM 2004), Nantes, pp. 59–66 (2004)
2. Beyer, H.G., Sendhoff, B.: Robust optimization-a comprehensive survey. Comput. Methods Appl. Mech. Eng. **196**(33), 3190–3218 (2007)
3. Chabin, T., Tonda, A., Lutton, E.: Is global sensitivity analysis useful to evolutionary computation? In: Proceedings of the Companion Publication of the 2015 on Genetic and Evolutionary Computation Conference, pp. 1365–1366. ACM (2015)
4. Collet, P., Lutton, E., Schoenauer, M., Louchet, J.: Take it easea. In: Deb, K., Rudolph, G., Lutton, E., Merelo, J.J., Schoenauer, M., Schwefel, H.-P., Yao, X. (eds.) PPSN 2000. LNCS, vol. 1917, pp. 891–901. Springer, Heidelberg (2000)

5. Deb, K., Jain, S.: Running performance metrics for evolutionary multi-objective optimizations. In: Proceedings of the Fourth Asia-Pacific Conference on Simulated Evolution and Learning (SEAL 2002), Singapore, pp. 13–20 (2002)
6. Deb, K., Pratap, A., Agarwal, S., Meyarivan, T.: A fast and elitist multiobjective genetic algorithm: NSGA-II. IEEE Trans. Evol. Comput. 6(2), 182–197 (2002)
7. Fu, G., Kapelan, Z., Reed, P.: Reducing the complexity of multiobjective water distribution system optimization through global sensitivity analysis. J. Water Resour. Plann. Manag. 138(3), 196–207 (2011)
8. Goldberg, D., Korb, B., Deb, K.: Messy genetic algorithms: motivation, analysis, and first results. Complex Syst. 3(5), 493–530 (1989)
9. Goldberg, D.: Genetic algorithms and walsh fuctions: II, deception and its analysis. Complex Syst. 3(2), 153–171 (1989)
10. Goldberg, D.: Genetic algorithms and walsh functions: I, a gentle introduction. Complex Syst. 3(2), 129–152 (1989)
11. Hansen, N., Ostermeier, A.: Completely derandomized self-adaptation in evolution strategies. Evol. Comput. 9(2), 159–195 (2001)
12. Kargupta, H.: The gene expression messy genetic algorithm. In: International Conference on Evolutionary Computation, pp. 814–819 (1996)
13. Leblanc, B., Lutton, E.: Bitwise regularity and ga-hardness. In: ICEC 1998, 5–9 May, Anchorage, Alaska (1998)
14. Lutton, E., Lévy Véhel, J.: Pointwise regularity of fitness landscapes and the performance of a simple ES. In: CEC 2006, Vancouver, Canada, 16–21 July 2006
15. Lutton, E., Véhel, J.L.: Hölder functions and deception of genetic algorithms. IEEE Trans. Evol. Comput. 2(2), 56–72 (1998)
16. Müller, C., Paul, G., Sbalzarini, I.: Sensitivities for free: CMA-ES based sensitivity analysis. Technical report, ETH Zurich (2011)
17. Paul, G., Müller, C., Sbalzarini, I.: Sensitivity analysis from evolutionary algorithm search paths. Technical report, ETH Zurich (2011)
18. Rajeev, S., Krishnamoorthy, C.: Genetic algorithms-based methodologies for design optimization of trusses. J. Struct. Eng. 123(3), 350–358 (1997)
19. Saltelli, A., Ratto, M., Andres, T., Campolongo, F., Cariboni, J., Gatelli, D., Saisana, M., Tarantola, S.: Global Sensitivity Analysis: The Primer. Wiley, New York (2008)
20. Saltelli, A., Annoni, P.: How to avoid a perfunctory sensitivity analysis. Environ. Model. Softw. 25(12), 1508–1517 (2010)
21. Sobol, I.M.: Global sensitivity indices for nonlinear mathematical models and their Monte Carlo estimates. Math. Comput. Simul. 55(1–3), 271–280 (2001)
22. Stonedahl, F., Wilensky, U.: Evolutionary robustness checking in the artificial anasazi model. In: AAAI Fall Symposium: Complex Adaptive Systems (2010)
23. Tang, Y., Reed, P., Wagener, T., Van Werkhoven, K., et al.: Comparing sensitivity analysis methods to advance lumped watershed model identification and evaluation. Hydrol. Earth Syst. Sci. Dis. 11(2), 793–817 (2007)
24. Zitzler, E., Thiele, L.: Multiobjective optimization using evolutionary algorithms - a comparative case study. In: Eiben, A.E., Bäck, T., Schoenauer, M., Schwefel, H.-P. (eds.) PPSN 1998. LNCS, vol. 1498, pp. 292–301. Springer, Heidelberg (1998)

Quasi-random Numbers Improve the CMA-ES on the BBOB Testbed

Olivier Teytaud$^{(\boxtimes)}$

TAO (Inria), LRI, UMR 8623 (CNRS - University Paris-Sud),
Bat 490 University Paris-Sud, 91405 Orsay, France
`teytaud@lri.fr`

Abstract. Pseudo-random numbers are usually a good enough approximation of random numbers in evolutionary algorithms. But quasi-random numbers follow a different idea, namely they are aimed at being more regularly distributed than random points. It has been pointed out in earlier papers that quasi-random points provide a significant improvement in evolutionary optimization. In this paper, we experiment quasi-random mutations on a well known test case, namely the Coco/Bbob test case. We also include experiments on translated or rescaled versions of BBOB, on which we get similar improvements.

1 Introduction

Monte Carlo is a classical method for computing approximate integrals. They can also be used directly for optimization; this is the simple random search algorithm. Evolutionary algorithms can be viewed as an improved form of random search, adaptively modifying the probability distribution in order to focus on the optimum. While Monte Carlo integration has been upgraded to Quasi Monte Carlo (also known as quasi-random), most evolution strategies use pseudo-random numbers, aimed at approximating random numbers, and not Quasi Monte Carlo, in spite of a few promising works in that direction. This might be due to lack of extensive experimental results on some classical testbeds; the purpose of this paper is to do this extensive experiment of quasi-random mutations in the Bbob/Coco benchmark. In this paper we recall the state of the art in the use of quasi Monte Carlo in evolution strategies (Sect. 2), and then experiment an existing quasi Monte Carlo evolutionary algorithm on the Bbob/Coco framework.

2 Derandomization in Evolution Strategies

Evolution strategies [1] have been "derandomized" in several manners: use of covariance matrix [2,6], and use of quasi-random points. We here consider the latter. It can be considered independently of the first and we will indeed use perform experiments in an algorithm which includes covariance matrix adaptation.

© Springer International Publishing Switzerland 2016
S. Bonnevay et al. (Eds.): EA 2015, LNCS 9554, pp. 58–70, 2016.
DOI: 10.1007/978-3-319-31471-6_5

Low-dispersion or quasi-random points have been used for derandomizing the random search [9,11,12], or evolutionary algorithms [8] or other randomized optimization algorithms [5]. We here refer mainly to [15,16], using quasi-random points for derandomizing the mutations in the CMA-ES algorithm [7]. The quasi-randomized version of CMA is termed DCMA, which stands for derandomized-CMA. Some important elements about quasi-random points follow. Computational cost is not a good reason for discarding quasi-random sequences. The computational cost for generating quasi-random points is negligible and indeed often smaller than for classical pseudo-random numbers [20,21]. Quasi-random sequences are different from pseudo-random sequences. Quasi-random numbers are not a special case of pseudo-random numbers. Pseudo-random sequences are aimed at imitating random sequences, whereas quasi-random sequences are aimed at doing better, thanks to a better uniformity. Additionally, modern quasi-random sequences have a random part [10]. Quasi-random points have low discrepancy, decreasing as the inverse of the number of points (within logarithmic factors), whereas pseudo-random numbers and random numbers, by design, have discrepancy decreasing as the inverse of the square root of the number of points. Pseudo-random numbers are an approximation of random numbers, whereas quasi-random numbers are qualitatively different. The weaknesses of old quasi-random sequences (such as non-scrambled Halton sequences), which were often worse than random sequences in high dimension, have been overcome thanks to randomized quasi-random sequences; these sequences have the good properties of quasi-Monte Carlo methods and are at least as performant as Monte Carlo methods in most (if not all) cases [13,14,17–19].

3   Experimental Results

We follow the experimental setup proposed in "exampleexperiment.m" provided in the Bbob/Coco downloads; a comment in the file states that the number of function evaluations should be increased, so we increase to $100 \times D$ with D the dimension for the strict Bbob/Coco setting in Sect. 3.1, which will be extended to $2000D$ in Sect. 3.3. We will also check translated or rescaled versions of Bbob. All experiments are performed with initial point $(0, 0, \ldots, 0)$ and initial step-size 1. The version of CMA-ES is the Matlab/Octave one as of the time of submission. All quasi-random numbers are obtained by the scrambled Halton method.

3.1   Experimental Results in the Bbob/Coco Setting

In this section, we produce results using the Bbob/Coco framework, without any change except the increase of the number of evaluations to $100 \times D$ (we increased this because it is recommended in the Bbob/Coco sample file to do so). The Bbob/Coco framework has been used in several conferences. Results are presented in Fig. 1 (frequency of success depending on the number of evaluations,

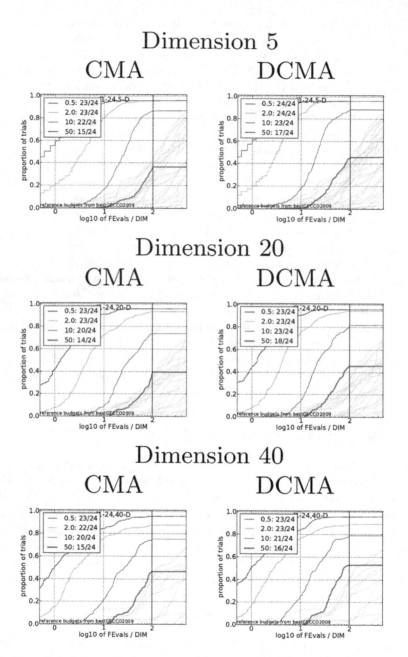

Fig. 1. Experimental results (percentage of success for different numbers of function evaluations; each curve corresponds to a different success criterion in terms of simple regret) for the default Bbob framework. Left: results with the default Cma. Right: results with DCMA. DCMA is usually faster. Figure 2 presents the same results as scatter plots.

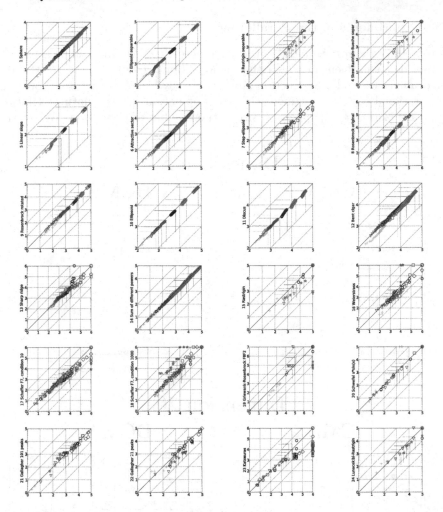

Fig. 2. Experimental results (scatterplots) for the default Bbob framework. For each graph, corresponding to one function from f1 to f24 in Bbob/Coco, the x-axis is the run length for the default CMA in log-10 scale, whereas the y-axis is the run length in log-10 scale for the quasi-randomized version, i.e. DCMA. CMA is better than DCMA for function f18, in the sense that there are more points above the diagonal than below. DCMA is better for the 22 other functions.

for different precision levels). Figure 2 presents the scatter plots, i.e. the x-axis is the computation time for reaching some precision for the default CMA whereas the y-axis is the computation time for reaching the same precision for DCMA. All graphs are obtained by Bbob/Coco automatically, so that there is no parameter choice by ourselves. All experiments use BBOB V13.09.

3.2   Experiments in the Parallel Setting

We reproduce the results above in the parallel setting. We will assume here that we consider a problem in which the computational cost is mainly in the fitness evaluations, and that function evaluations have an approximately constant computational cost, so that increasing the population size is a natural solution for parallelization: the population size is the number of processors. We set the

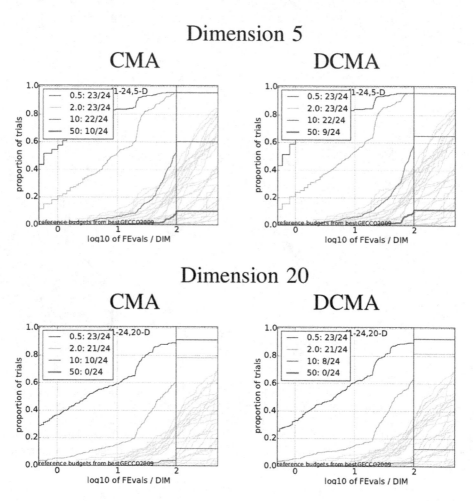

Fig. 3. Experiments on Bbob with population size forced to a larger value $20 \times D$ where D is the dimension. Success rates for different number of function evaluations as in Fig. 1. Left: results with the default CMA. Right: results with quasi-randomization (DCMA). Results are usually better for DCMA, but the difference is smaller than with the standard population size of CMA. Y-axis is a frequency of success: the higher the better.

population size to $20 \times D$, where D is the dimension, and do not modify anything else in the Bbob/Coco framework. Results are presented in Fig. 3 (frequency of approximate solving on the y-axis for the number of evaluation given on the x-axis).

3.3 Experiments with Larger Numbers of Iterations

We come back to the original Bbob/Coco setting of Sect. 3.1, but with $2000 \times D$ function evaluations in dimension D. Results are presented in Fig. 4 and still

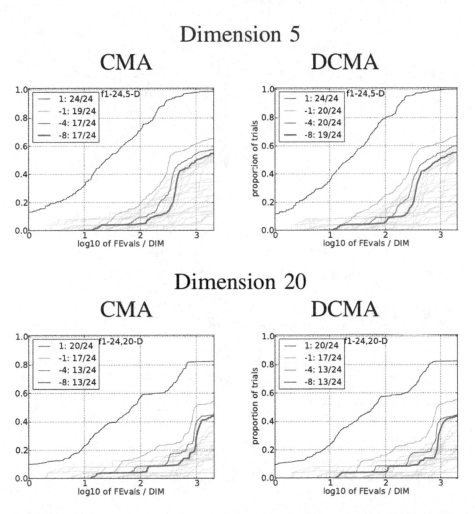

Fig. 4. Results in the original Bbob setting but with larger numbers ($2000D$) of function evaluations. Left: results with the default Cma. Right: results with quasi-randomization. The difference is smaller than in the other cases. Y-axis is still the frequency of success (the higher, the better)

show a superiority of DCMA but with a smaller difference. Detailed results show a strong superiority for f12, f15, f16, f17, f18, f19, f23, f24.

3.4   Experiments with Large Population Size and Large Numbers of Iterations

We come back to the Bbob/Coco setting of Sect. 3.2, i.e. population size equal to $20D$ where D is the dimension, but with $10000 \times D$ function evaluations in dimension D. Results are presented in Figs. 5 and 6 and still show a superiority of DCMA, though not for all functions.

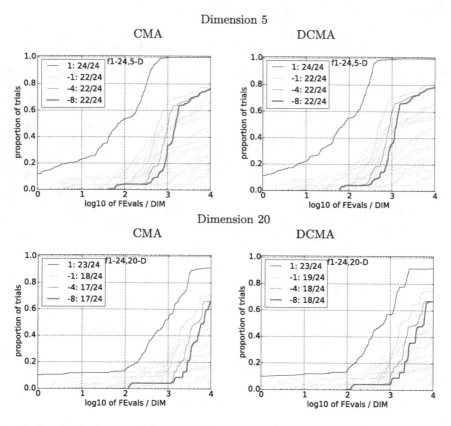

Fig. 5. Results in the parallel setting (20D as population) and with larger numbers of function evaluations (10000D). Left: results with the default Cma. Right: results with quasi-randomization. Results similar to the standard case.

4   Experiments with Modified BBOB

In this section, we rescale the BBOB testbed. As in the original experiments (Sect. 3.1), we use $100D$ function evaluations. Instead of working on $f(x)$, we

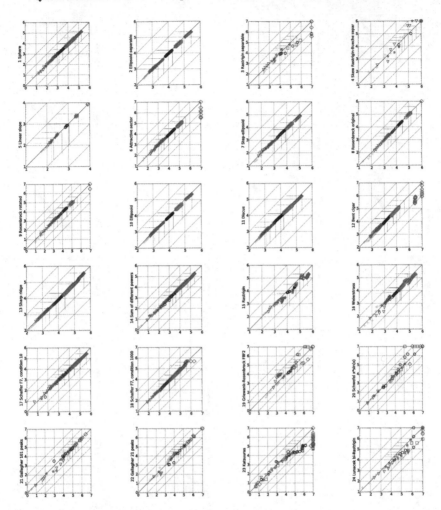

Fig. 6. Scatterplots in the parallel setting as in Sect. 3.2 (population size $20D$) but with larger numbers of function evaluations ($10000D$); text in Sect. 3.4. For each graph, corresponding to functions f1 to f24 in the Bbob/Coco framework, the x-axis is the run length for the default CMA in log-10 scale, whereas the y-axis is the run length in log-10 scale for the quasi-randomized version. DCMA outperforms CMA in the sense that there are more points below the diagonal than above for most functions, but the difference is often small; the difference is bigger for f3, f6, f12, f15, f16, f23, f24. CMA outperforms DCMA for f4 and f20.

work on $f(x/1000)$. Results are presented in Fig. 7. The superiority of DCMA over CMA is bigger, suggesting that derandomized mutations improve the robustness w.r.t an imperfect initialization (guessing the initial step-size is not that easy in real situations) leads to a roughly linear landscape.

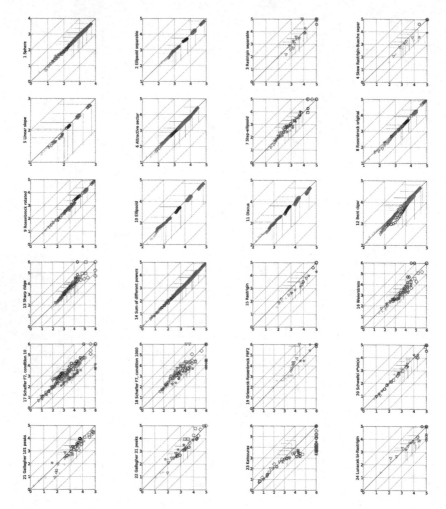

Fig. 7. Comparison between CMA and DCMA on the rescaled testbed. The difference between CMA and DCMA is similar to the difference in the original BBOB testbed. The improvement is visible on nearly all functions except f19 in the sense that there are more points below than above the curve. The difference is clearer on multimodal functions.

5 Experiments with Another Modified BBOB

In this section, we translate the BBOB testbed. As in the original experiments (Sect. 3.1), we use $100D$ function evaluations. Instead of working on $f(x)$, we work on $f(x + 7)$ ($+7$ is added coordinate-wise, i.e. all d decision variables are shifted in dimension d). Results are presented in Fig. 8. The improvement by DCMA over CMA is bigger than in the original BBOB.

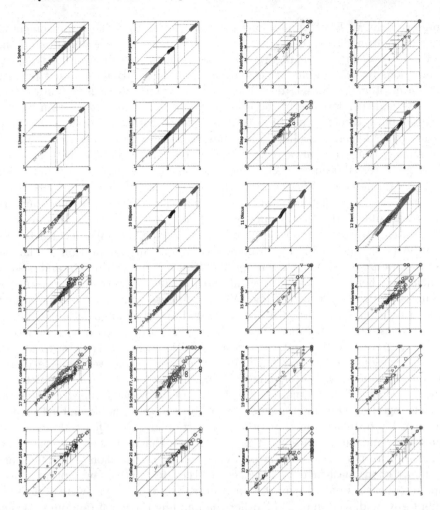

Fig. 8. Comparison between CMA and DCMA on the translated testbed. Each graph represents a function. X-axis = number of function evaluations for reaching the target precision for CMA. Y-axis = number of function evaluations for reaching the target precision for DCMA. The difference between CMA and DCMA is similar to the difference in the original BBOB testbed. The improvement is visible on nearly all functions (in the sense that we have more points below than above the curve), in particular multimodal.

6   Conclusion

The derandomization proposed in [15] basically works. There are settings in which the difference is large, and settings in which the effect of quasi-randomization is minor; but it is rarely detrimental. The contribution of this paper are (i) confirming this superiority on the BBOB testbed (ii) efficiency of DCMA compared to CMA is preserved with large population sizes (iii) it is preserved in all Bbob dimensions (v) we confirm that the improvement is better in multimodal settings; this is consistent with [3]. We perturbated the BBOB testcase, just by changing the scale by a factor 1000, or by translating functions by +7. Results are essentially preserved. BBOB does not provide confidence intervals. This is deeply rooted in BBOB: there is a finite set of functions, and therefore overfitting is always possible, a trivial algorithm successively sampling the finite set of optima of BBOB instances for the considered dimension would have excellent performance. Nonetheless, we reproduced the results many times, and always got the same result, including translations and rescaling. All tested frameworks have been presented. We considered results with respect to the number of fitness evaluations, not computation time; this is the standard Coco/Bbob methodology. The computational cost of the quasi-random part is negligible, indeed the computational complexity of quasi-random numbers is often less than the one of pseudo-random numbers. We decided to run experiments on the Bbob/Coco framework without any adaptation so that at least the framework is not chosen specifically for the experiments and results are neutral. There was no tuning at all and presented results are the results of the first set of runs in each setting. We now discuss limitations of the present paper. In the present work, we just validated the derandomization of mutations by quasi-random numbers. Other derandomizations, based on symmetries as the one proposed in [4], might provide additional improvements; these two derandomizations can be combined. In the present paper we combine quasi-Monte Carlo and Covariance Matrix Adaptation, we could have symmetrized sampling combined with quasi-Monte Carlo and Covariance Matrix Adaptation, all together. Our experiments are performed with the scrambled Halton sequence. We do not claim that other, in particular older Quasi-Monte Carlo sequences would be as efficient. It is well known that old Quasi Monte Carlo sequences were not that good, in particular in high dimension [13]. There are now many good quasi-random sequences in the literature. Maybe other quasi-random sequences would provide better results. The experiments were performed without any modification of CMA other than adding the quasi-random part, i.e. replacing $arz = random\ gaussian$ by $arz = quasi\ random\ gaussian$ (where arz is the notation in CMA for the mutation before rescaling and applying the covariance transformation). It is likely that the optimal parameters for the covariance update and for the step-size update are different from the optimal parameters for the original CMA. Therefore there is likely margin for improving the results of the DCMA algorithm, which is left as further work. Quasi-random, or low-dispersion, can be used also for the restarts. This is the purpose of other published works. We did not include quasi-random restarts in order to separate both effects. Still, the performance improvement might be due to a better spreading of the initialization over the domain. We conjecture that the improvement related to quasi-random restarts will be larger than the one with quasi-random mutations - the purpose of this paper is basically that we can also include quasi-randomization in mutations.

References

1. Beyer, H.G.: The Theory of Evolution Strategies. Natural Computing Series. Springer, Heideberg (2001)
2. Beyer, H.-G., Sendhoff, B.: Covariance matrix adaptation revisited – the CMSA evolution strategy. In: Rudolph, G., Jansen, T., Lucas, S., Poloni, C., Beume, N. (eds.) PPSN 2008. LNCS, vol. 5199, pp. 123–132. springer, Heidelberg (2008)
3. Chaslot, G., Hoock, J.B., Teytaud, F., Teytaud, O.: On the huge benefit of quasi-random mutations for multimodal optimization with application to grid-based tuning of neurocontrollers. In: ESANN (2009). http://dblp.uni-trier.de/db/conf/esann/esann2009.html#ChaslotHTT09
4. Teytaud, O., Gelly, S., Mary, J.: On the ultimate convergence rates for isotropic algorithms and the best choices among various forms of isotropy. In: Runarsson, T.P., Beyer, H.-G., Burke, E.K., Merelo-Guervós, J.J., Whitley, L.D., Yao, X. (eds.) PPSN 2006. LNCS, vol. 4193, pp. 32–41. Springer, Heidelberg (2006)
5. Georgieva, A., Jordanov, I.: A hybrid meta-heuristic for global optimisation using low-discrepancy sequences of points. Comput. Oper. Res. Spec. Issue Hybrid Meta-heuristics **37**, 429 (2010)
6. Hansen, N., Ostermeier, A.: Adapting arbitrary normal mutation distributions in evolution strategies: the covariance matrix adaption. In: Proceedings of the IEEE Conference on Evolutionary Computation (CEC 1996), pp. 312–317. IEEE Press (1996)
7. Hansen, N., Ostermeier, A.: Completely derandomized self-adaptation in evolution strategies. Evol. Comput. **11**(1), 1 (2003)
8. Kimura, S., Matsumura, K.: Genetic algorithms using low-discrepancy sequences. In: GECCO, pp. 1341–1346 (2005)
9. Lindemann, S.R., LaValle, S.M.: Incremental low-discrepancy lattice methods for motion planning. In: Proceedings IEEE International Conference on Robotics and Automation, pp. 2920–2927 (2003)
10. Mascagni, M., Chi, H.: On the scrambled halton sequence. Monte-Carlo Methods Appl. **10**(3), 435–442 (2004)
11. Niederreiter, H.: Random Number Generation and Quasi-Monte-Carlo Methods. Society of Industrial and Applied Mathematics, Philadelphia (1992)
12. Niederreiter, H.: Low-discrepancy and low-dispersion sequences. J. Number Theor. **30**, 51 (1988)
13. Owen, A.: Multidimensional variation for quasi-Monte-Carlo (2004)
14. Sobol, I.M.: On the systematic search in a hypercube. SIAM J. Numer. Anal. **16**(5), 790–793 (1979)
15. Teytaud, O.: When does quasi-random work? In: Rudolph, G., Jansen, T., Lucas, S., Poloni, C., Beume, N. (eds.) PPSN 2008. LNCS, vol. 5199, pp. 325–336. Springer, Heidelberg (2008). http://dblp.uni-trier.de/db/conf/ppsn/ppsn2008.html#Teytaud08
16. Teytaud, O., Gelly, S.: DCMA, yet another derandomization in covariance-matrix-adaptation. In: D. Thierens et al. (ed.) GECCO, pp. 955–922. London Royaume-Uni (2007). http://hal.inria.fr/inria-00173207/en/
17. Tuffin, B.: A new permutation choice in halton sequences. In: Niederreiter, H., Hellekalek, P., Larcher, G., Zinterhof, P., et al. (eds.) Monte Carlo and Quasi-Monte Carlo Methods 1996. Lecture Notes in Statistics, vol. 127, pp. 427–435. Springer, New York (1997)

18. Vandewoestyne, B., Cools, R.: Good permutations for deterministic scrambled halton sequences in terms of l2-discrepancy. Comput. Appl. Math. **189**(1, 2), 341–361 (2006)
19. Wang, X., Hickernell, F.: Randomized halton sequences. Math. Comput. Model. **32**, 887–899 (2000)
20. Warnock, T.: Computational investigations of low-discrepancy point sets. In: Zaremba, S.K. (ed.) Applications of Number Theory to Numerical Analysis (Proceedings of the Symposium), University of Montreal, pp. 319–343 (1972)
21. Warnock, T.T.: Computational investigations of low-discrepancy point sets II. In: Niederreiter, H., Shiue, P.J.-S. (eds.) Monte Carlo and Quasi-Monte Carlo Methods in Scientific Computing. Lecture Notes in Statistics, vol. 106, pp. 354–361. Springer, New York (1995)

Progressive Differential Evolution on Clustering Real World Problems

Vincent Berthier[(✉)]

TAO (Inria), LRI, UMR 8623 (CNRS - Univ. Paris-Sud), Bat 660 Claude Shannon
Univ. Paris-Sud, 91190 Gif-sur-Yvette, France
`vincent.berthier@inria.fr`

Abstract. In this paper, we assess the performances of Differential Evolution on real-world clustering problems. To improve our results, we introduce Progressive Differential Evolution, a small modification of Differential Evolution which aims at optimizing a small number of parameters (*eg.* one cluster) at the beginning, and incrementally increase the number of optimized parameters.

1 Introduction

While many benchmarks used in the optimisation community to evaluate algorithms are based on purely artificial functions such as [10, 20], it can only be the first step in what ultimately is aimed at solving real world problems. Some recent initiatives went in that direction (see [8] for example), proposing new ways to assess the performances of optimisation algorithms.

In this paper, by comparing our results on one such benchmark, we (i) show that the Differential Evolution algorithm is very efficient on clustering problems and (ii) propose Progressive Differential Evolution, which starts with a low number of parameters to optimise and gradually increases it.

Section 2 describes the benchmark we used to compare our results to other algorithms and Sect. 3 validates our approach. Section 4 recalls the Differential Evolution algorithm and the "DE/curr-to-best/1" variant we used while Sect. 5 introduces Progressive Differential Evolution. In Sect. 6 we compare our results to the state of the art.

2 Continuous Real-World Representative Benchmark

Most of the existing testbeds used to evaluate optimisation algorithm compare their performances on artificial functions, such as the sphere, the ellipsoid or the Rosenbrock function to cite the most notable ones. With the improvements of the algorithms, more complex functions were introduced with some specific properties such as rotation, non separability, multimodality and so on, but ultimately, most testbeds are completely artificial.

While this is by no mean uninteresting, the ultimate goal in optimisation is to solve real world problems. The gap between artificial functions - as complex

S. Bonnevay et al. (Eds.): EA 2015, LNCS 9554, pp. 71–82, 2016.
DOI: 10.1007/978-3-319-31471-6_6

as they are - to real world issues seems too large to directly apply what we know. As such, new testbeds, with some real world properties are advisable.

One of such propositions comes from [7] and revolves around clustering problems that have interesting properties to evaluate optimisation algorithms: challenging, scalable, easy to understand and implement, and most of all, their data can - and should - come from real world examples. Each cluster is used as a vector of coordinates in the parameters' space of data, which allows us to use optimisation algorithms on those problems.

The three problems used here are the Iris [6], the Ruspini [15] and the German Town [18] datasets, all of them widely used in the clustering community to evaluate the performances of their own algorithms, and rooted in the real world. More importantly, [12] computed the global optimum for those datasets from two to ten clusters, which allows us to assess the performances of the algorithms. The German Town points are defined in 3D, the Iris ones in 4D and for Ruspini it is in 2D.

Along with a k-means clustering algorithm, [7] studied the performances of three black-box algorithms: CMA-ES [9] (one with standard population size, one with an increased population), Nelder-Mead [13] and Random-Search. One of the conclusions is that even if the k-means algorithm converges very quickly, it is often beaten by CMA-ES (with increased population size) in term of quality of the solution found. Thus, complete black-box algorithms are able to outperform problem specific ones.

3 Implementation Validation

In order to compare results obtained on our platform using Evolving Objects (see [11]), we ran the benchmark on two CMA-ES with the same configuration as [7]: one has default parameters, one has a population size of $\mu = 50$ and $\lambda = 100$. In both cases, we stopped a run when $f_{best} \leq f^* + \frac{f^*}{1e15}$ (*ie.* the optimum is reached), when the best fitness stagnated for too long or when the allocated budget was consumed. This budget was set to $2e5$ function evaluations (all budgets in this paper are expressed in terms of function evaluations).

As can be seen in Table 1, the mean fitnesses we were able to obtain are comparable to the ones reported in [7]: sometimes better, sometimes worse, but never by far (except in high dimension where the results are degraded, probably due to different parameters). This allows us to validate our implementation, and serves as a baseline for the rest of our work.

In the original paper, the number of function evaluations was reported with the mean fitnesses. The given explanation is that the main focus of the exercise being the fitness - and not so much failures or successes - the required number of function evaluations to get a result is not that important: each algorithm should have the time - the budget - to reach the optimum or at least converge.

While this is perfectly valid, we don't feel comfortable to do so as it weakens the comparison between algorithms. Instead of reporting the mean number of function evaluations used, we will prefer the SP1 measure as defined in [1]:

Table 1. Average fitness results and SP1 measure (mean and standard deviation) for CMA-ES and CMA-ES(50,100). An SP1 measure of ∞ means that the optimum could not be reached for any of the 50 runs. Results are give for the German Town (G), Iris (I) and Ruspini (R) datasets for all values of k.

D	k	f*	CMA-ES (50,100)	CMA-ES (50,100) SP1	CMA-ES	CMA-ES SP1
G	02	6.02546e11	6.025472e11 (2.8e-04)	8.3400e03 (6.4e02)	1.172558e12 (7.6e11)	4.8798e04 (3.3e04)
	03	2.94506e11	4.486461e11 (1.5e11)	4.2083e04 (2.6e04)	8.196432e11 (1.2e12)	∞
	04	1.04474e11	3.362127e11 (1.4e11)	3.9970e05 (1.8e05)	7.629370e11 (4.7e11)	∞
	05	5.97615e10	2.049802e11 (1.4e11)	6.8410e05 (2.0e05)	7.488858e11 (1.2e12)	∞
	06	3.59085e10	1.585765e11 (1.5e11)	∞	8.818792e11 (6.3e11)	∞
	07	2.19832e10	1.051648e11 (1.1e11)	∞	6.463187e11 (7.5e11)	∞
	08	1.33854e10	1.068587e11 (9.3e10)	∞	7.005948e11 (7.1e11)	∞
	09	7.80442e09	2.780667e11 (3.1e11)	∞	1.003192e12 (9.7e11)	∞
	10	6.44647e09	5.869352e11 (5.2e11)	∞	7.677317e11 (6.5e11)	∞
I	02	1.52348e2	1.523480e02 (6.4e-14)	1.8344e04 (5.0e02)	1.523542e02 (3.0e-03)	∞
	03	7.88514e01	7.885144e01 (2.5e-14)	7.2048e04 (2.2e03)	1.279512e02 (1.2e02)	∞
	04	5.72285e01	5.836730e01 (3.9e00)	∞	9.728522e01 (3.5e01)	∞
	05	4.64462e01	4.766177e01 (1.7e00)	∞	1.330878e02 (1.3e02)	∞
	06	3.90400e01	4.149195e01 (2.9e00)	∞	1.292478e02 (1.3e02)	∞
	07	3.42982e01	4.037920e01 (3.5e00)	∞	7.892632e01 (4.5e01)	∞
	08	2.99889e01	3.739813e01 (4.2e00)	∞	7.750688e01 (5.4e01)	∞
	09	2.77861e01	3.831817e01 (5.3e00)	∞	8.018775e01 (7.6e01)	∞
	10	2.58341e01	5.653196e01 (6.9e01)	∞	9.553900e01 (1.0e02)	∞
R	02	8.93378e04	8.933783e04 (5.0e-12)	6.8260e03 (1.1e03)	8.933783e04 (3.1e-11)	3.5903e04 (5.0e03)
	03	5.10635e04	5.110393e04 (4.6e01)	2.0453e04 (5.3e03)	5.473043e04 (9.8e03)	∞
	04	1.28811e04	1.288105e04 (0.0e00)	∞	2.046652e04 (1.5e04)	∞
	05	1.01267e04	1.032449e04 (5.0e02)	∞	3.209521e04 (1.4e04)	∞
	06	8.57541e03	8.919118e03 (5.1e02)	2.5490e05 (1.7e04)	2.605724e04 (1.3e04)	∞
	07	7.12620e03	7.634386e03 (4.4e02)	7.7641e05 (4.9e04)	2.309534e04 (6.1e03)	∞
	08	6.14964e03	6.635902e03 (3.9e02)	∞	2.061007c04 (5.2e03)	∞
	09	5.18165e03	7.464273e03 (3.6e03)	∞	1.906988e04 (5.3e03)	∞
	10	4.44628e03	1.095691e04 (5.0e03)	∞	1.696298e04 (5.6e03)	∞

$SP1 = \frac{\mathbb{E}(T_s)}{p_s}$, where $\mathbb{E}(T_s)$ is the expected number of function evaluations used in a successful run and p_s is the probability to get a success for a given run.

This measure has some disadvantages (*eg.* when the success probability is 0), but it allows a more accurate comparison between algorithms, in particular when using restarts. In such a way, two possible strategies (aiming for a 100 % success rate no matter the cost or allowing restarts if the solution is not quickly found) are both possible and their performances can be compared without bias one way or another.

4   Differential Evolution

While the first work on this clustering benchmark obviously did not try to compare each and every possible optimisation algorithm, we felt that given the specifities of the problem, Differential Evolution (DE) [19] could perform quite well. This feeling is substantiated by [4] in which DE is said to perform very well on a lot of testbeds.

Built around crossovers, the DE algorithm replaces part of a given individual with two or more others. Many different variants of DE exist, each one defining the crossovers rule. The one we chose was "DE/curr-to-best/1". For a given generation, we then have:

DE/curr-to-best/1: $U(0, 1)$ is a random uniformly distributed number between 0 and 1, CR is the crossover rate parameter, f_1 and f_2 are two real numbers, $Best$ is the best individual in the generation, and f is the evaluation function, n is the dimension of a point in the given dataset.

for each individual I **do**
   $Y \leftarrow I$
   Randomly choose A and B, two individuals distinct from I and $Best$
   Randomly select an index $R \in \{1, \ldots, n\}$
   for all $i \in \{1, \ldots, n\}$ **do**
      if $i = R$ **or** $U(0, 1) < CR$ **then**
         $Y(i) \leftarrow I(i) + f_1(A(i) - B(i)) + f_2(Best(i) - I(i))$
      end if
   end for
   if $f(Y) < f(I)$ **then**
      Replace I by Y
   end if
end for

The only difference from "DE/best/1" is thus the update formula, which is $Y(i) \leftarrow Best(i) + f_1(A(i) - B(i))$.

In the spirit of [7], we didn't try to tune the algorithm's parameters. Instead, in the absence of a standard recommendation, we set $CR = 0.5$, $f_1 = f_2 = 0.8$ for a population size of 30. The initialisation points were randomly drawn with a normal distribution of mean the average of the range of the variables, with a standard deviation of a third of that average. We used here the same stopping criteria as with CMA-ES in our previous experiment.

5 Progressive Differential Evolution

In some of our first trials, when studying the reasons for failures to reach the optimum, we reached the conclusion that in a third of the failed runs, this failure was due to falling in a local optimum. As can be seen on Fig. 1 with a 3e4 budget, in most cases the failures to reach the optimum are simply due to a lack of budget: the clusters found are not exactly at the optimum but centered around them. In fact, by increasing the budget, we saw that indeed, those points went to the optimum.

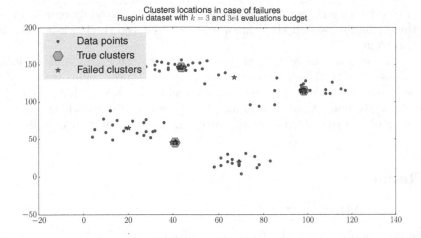

Fig. 1. Clusters position on failure cases, Ruspini dataset with $k = 3$

In the second case however, we can see that the points found are symmetrically opposed to the optimum solution, one cluster at the top, two at the bottom. This configuration on the Ruspini problem with $k = 3$ gives a fitness of ≈ 51155, which is only slightly worse than the optimum of ≈ 51063. As such, there is only a very small probability that any mutation would get to the real optimum close enough to improve the solution.

In order to avoid this, we introduced "Progressive Widening", known as the Sieves Method [16] in statistics. The basic idea is to start optimising a small number of clusters, and to increase that number at some point in the process:

PDE: k_{max} is the desired number of clusters, N is the dimensionality of each point, R determines the number of generations to do with k clusters

   Initialise population
   $k \leftarrow 1$
   while not stop **do**
      for $i = 0$ to R **do**
         Run one generation of DE on the $k \cdot N$ first parameters
      end for
      $k \leftarrow min(k + 1, k_{max})$
   end while

Here, we chose to use $R = 100$, which means that every one hundred generation, we increase the number of parameters to optimise until we reach $n \cdot k_{max}$.

Of course, the fact that we optimise k clusters doesn't mean that the others "disappear": they are still taken into account in the evaluation, but don't move from their initial position, which is the center of the search space. This means that even when training k clusters, there is always one more that can be selected as the nearest from a given point. While we could have completely removed them from the evaluation, we felt that this would have reduced the black-box context of the problem.

In fact, one could argue that we are only able to use Progressive Widening by weakening the black-box setting of the problem. Indeed, since we know the dimension of the problem, we know that to add a cluster we have to add N parameters. We don't think this is an issue however, this knowledge being as much part of the specification of the problem as the definition of the search space.

6 Results

6.1 De vs CMA-ES

The results we obtained with DE shown in Table 2 and Fig. 3 were very good, often better - sometimes by far - than CMA-ES(50,100). The first striking result is that DE more consistently reaches the optimum solution: in only five cases (three on the German Town dataset, two on the Ruspini dataset) DE was not able to reach the optimum at least once in the 50 runs reported here.

As such, it comes as no surprise that the average fitness obtained by DE after 50 runs was improved in almost all cases (except on the Iris dataset when $k \leq 6$ and on the Ruspini dataset with $k = 3$). While this improvement is not necessarily ground breaking on the Ruspini dataset for example, it is much more important on the German Town problem (see Fig. 3a).

6.2 De vs PDE

The effects of the Progressive Widening on DE were twofold: first, it globally improved the average fitness across the board: in all but one trial (Ruspini with $k = 7$), the mean fitness and associated standard deviation were better with Progressive Widening than without. Once more, this is most notable on the German Town problem. Furthermore, in only one case now (Iris dataset with $k = 10$) is CMA-ES the best: on all other cases, PDE gets better results.

The second effect (shown in Table 3) was the one we expected: the success rate improved, we find the optimum more often. Most notably, with $k = 3$ on the Ruspini dataset, we went up from a 62 % success rate to a full 100 %: we no longer fall in the local optimum reported in Fig. 1, which was our goal when adding Progressive Widening to DE.

In five cases though the rates went down but only in two cases was this decrease important: from 58 % to 20 % on the Ruspini dataset with $k = 7$ (which is also the only case where the mean fitness obtained by DE is better than PDE) and from 32 % to 6 % still on the Ruspini dataset but with $k = 10$. Interestingly here, while the success rate decreased by almost 30 %, the mean fitness obtained by PDE is still better than the one from DE.

In fact thanks to this, we can see that while the Progressive Widening works very well in most instances in order to avoid a local minimum, in some rare

Table 2. Average fitness results and SP1 measure (mean and standard deviation) for DE and PDE. An SP1 measure of ∞ means that the optimum could not be reach for any of the 50 runs. Results are give for the German Town (G), Iris (I) and Ruspini (R) datasets for all values of k.

D	k	$f*$	DE	DE SP1	PDE	PDE SP1
G	02	6.02546e11	6.025472e11 (5.0e-04)	5.6160e03 (6.6e02)	6.025472e11 (5.0e-04)	9.1242e03 (2.9e02)
	03	2.94506e11	3.006674e11 (4.3e10)	9.0893e03 (1.8e03)	2.945066e11 (0.0e00)	1.3496e04 (2.6e02)
	04	1.04474e11	1.500823e11 (8.1e10)	2.3898e04 (1.6e04)	1.044747e11 (0.0e00)	1.9849e04 (4.5e02)
	05	5.97615e10	7.423346e10 (3.6e10)	4.5462e04 (4.3e04)	6.065579e10 (6.3e09)	2.7124e04 (1.6e03)
	06	3.59085e10	4.776401e10 (3.8e10)	4.8107e04 (2.3e04)	3.611288e10 (1.4e09)	3.3399e04 (1.5e03)
	07	2.19832e10	3.176165e10 (1.6e10)	2.0357e05 (1.1e05)	2.423709e10 (5.1e09)	8.8896e04 (3.5e04)
	08	1.33854e10	2.182272e10 (8.3e09)	∞	1.639762e10 (4.1e09)	∞
	09	7.80442e09	1.562879e10 (6.3e09)	∞	1.127751e10 (2.9e09)	∞
	10	6.44647e09	1.281459e10 (6.2e09)	∞	8.793075e09 (4.3e09)	2.2032e06 (3.4e05)
I	02	1.52348e02	1.523480e02 (0.0e00)	8.9892e03 (2.7e03)	1.523480e02 (0.0e00)	1.3222e04 (3.5e02)
	03	7.88514e01	8.032188e01 (1.0e01)	2.0023e04 (1.3e04)	7.885212e01 (1.5e-03)	2.1965e04 (7.9e02)
	04	5.72285e01	5.867260e01 (5.2e00)	5.0221e04 (2.7e04)	5.722847e01 (4.8e-14)	2.6411e04 (5.6e03)
	05	4.64462e01	4.978281e01 (4.3e00)	1.3932e05 (8.1e04)	4.847058e01 (1.8e00)	1.7655e05 (5.8e04)
	06	3.90400e01	4.210588e01 (3.4e00)	1.9240e05 (7.7e04)	3.961203e01 (1.5e00)	1.2713e05 (4.7e04)
	07	3.42982e01	3.735682e01 (3.4e00)	1.3620e06 (3.7e05)	3.506627e01 (1.6e00)	4.1865e06 (0.0e00)
	08	2.99889e01	3.288639e01 (3.3e00)	1.0018e06 (1.2e05)	3.084912e01 (1.4e00)	9.5156e05 (1.3e05)
	09	2.77861e01	2.928749e01 (2.2e00)	1.6612e06 (1.1e05)	2.855921e01 (1.2e00)	1.3116e06 (2.8e05)
	10	2.58341e01	2.795759e01 (2.7e00)	3.5490e06 (0.0e00)	2.695760e01 (9.1e-01)	7.8765e06 (0.0e00)
R	02	8.93378e04	8.933783e04 (0.0e00)	5.9892e03 (2.3e03)	8.933783e04 (0.0e00)	1.0646e04 (4.0e02)
	03	5.10635e04	5.109841e04 (4.5e01)	2.0758e04 (9.1e03)	5.106348e04 (4.1e-11)	1.1740e04 (4.0e02)
	04	1.28811e04	1.288105e04 (0.0e00)	∞	1.288105e04 (0.0e00)	1.1175e06 (0.0e00)
	05	1.01267e04	1.015393e04 (1.9e02)	∞	1.013935e04 (1.1e01)	∞
	06	8.57541e03	8.664380e03 (2.5e02)	1.0505e05 (8.3e04)	8.660781e03 (1.1e02)	9.7329e04 (2.2e04)
	07	7.12620e03	7.179452e03 (1.4e02)	1.0829e05 (7.4e04)	7.193774e03 (1.0e02)	1.6640e05 (1.0e04)
	08	6.14964e03	6.246995e03 (3.6e02)	1.7645e05 (1.2e05)	6.168576e03 (3.4e01)	7.7184e04 (2.0e04)
	09	5.18165e03	5.441820e03 (4.2e02)	2.8236e05 (1.3e05)	5.314655e03 (1.9e02)	1.3664e05 (5.8e04)
	10	4.44628e03	4.694111e03 (4.3e02)	2.6013e05 (9.9e04)	4.622832e03 (8.4e01)	8.9633e05 (4.6e04)

Table 3. Success rate for CMA(50,100), DE and PDE

(a) German Town dataset

k	CMA(50,100)	DE	PDE
02	100	100	100
03	48	98	100
04	10	76	100
05	18	74	98
06	0	74	88
07	0	38	74
08	0	0	0
09	0	0	0
10	0	0	8

(b) Iris dataset

k	CMA(50,100)	DE	PDE
02	100	100	100
03	100	86	84
04	0	56	100
05	0	28	28
06	0	32	50
07	0	4	2
08	0	6	8
09	0	4	8
10	0	2	2

(c) Ruspini dataset

k	CMA(50,100)	DE	PDE
02	100	100	100
03	56	62	100
04	0	0	2
05	0	0	0
06	24	46	36
07	16	58	20
08	0	42	64
09	0	32	52
10	0	32	6

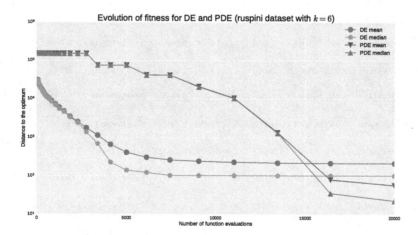

Fig. 2. Fitness statistics evolution on the Ruspini dataset with $k = 6$ with DE and PDE. The Progressive Widening has a clear cost at the beginning of the optimisation process.

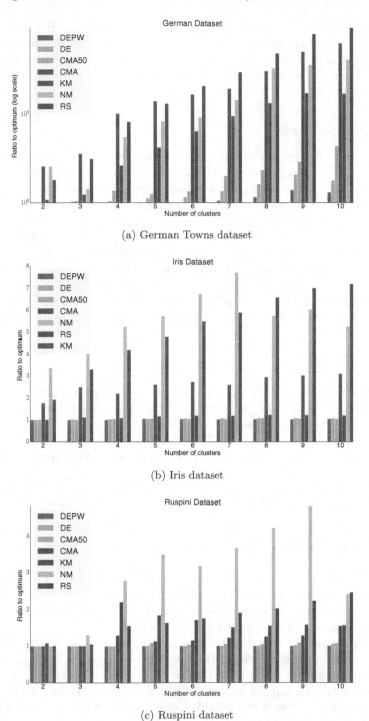

(a) German Towns dataset

(b) Iris dataset

(c) Ruspini dataset

Fig. 3. Performance as a ratio to the optimum ($\frac{\hat{f}}{f^*}$) of results reported in the original paper compared to DE and PDE with a $2e5$ budget. From left to right are PDE, DE, CMA-ES(50,100), CMA-ES, KM, NM and RS.

cases it is exactly the opposite, as we can see on Ruspini with $k = 10$. While the solution found is often very good - there is not a huge difference between DE and PDE mean fitness there - by plotting the proposed solution we see that when PDE fails to reach the optimum and stagnates, it is because it fell in a local minimum.

6.3   The Cost of PDE

Given the fact that the budget and stopping criteria are the same for DE and PDE, the SP1 measures reported in Table 2 mostly reflect the differences in success rate we saw previously. In the few cases were both algorithm have (almost) the same success rate, we can see that the SP1 measure is higher (or worse) for PDE than for DE: the introduction of Progressive Widening is not without cost.

This is even more clearly illustrated in Fig. 2, where some statistics on the fitnesses of 50 runs of DE and PDE are plotted. On the first few evaluations, PDE performs two orders of magnitude worse than DE, still one order of magnitude worse after $5e3$ evaluations, and it is not until at least $1.5e4$ evaluations that PDE performs at least as well as DE. While this is to be expected since until then not all clusters are optimised, it is still something to take into account.

7   Conclusion

DE performs very well on clustering problems, even when compared to clustering algorithms or CMA-ES, the current state of the art on this benchmark. This, by itself, is a very impressive result.

Our proposed variant of DE, PDE, gets even better results in most cases illustrating the good impact the concept of Progressive Widening can have on a black box algorithm.

In addition, we propose a baseline for the SP1 measure that will allow more robust comparisons of algorithms on this benchmark in the future.

8   Further Work

While still following the spirit of the original paper by not tuning the algorithms parameters, there are still many possibilities to try and improve the results. Some ways to do so include other mutations rules for DE (DE/rand/1, DE/best/1, *etc.*), using Adaptive Differential Evolution, or other variants.

Of course, another way could be to use the progressive strategy on other algorithms when possible: for algorithms with covariance matrices such as CMA-ES, CMSA [3] or even the self-adaptive with covariance algorithm [14] such a change is not trivial. But for others like Particle Swarm Optimisation [5,17] or the other members of the Self-Adaptive family [2] (isotropic or anisotropic, 1+1, *etc.*) this is quite straightforward.

The most interesting improvements could be done on the Progressive Widening concept. For example, knowing why in some instances it is more prone to fall in a local minimum would be interesting.

Furthermore, we have seen that the Progressive Widening is not without cost. To lessen that cost, instead of adding clusters (or parameters in the general case) at fixed timesteps we could design a rule that dynamically adds them when the fitness is reasonably stable. An intermediate step might be to add those parameters after an increasing number of timesteps (evaluations or generations) with a logarithmic rule for example, such that the more parameters are currently optimised, the more time is spent on them before adding more.

References

1. Auger, A., Hansen, N.: Performance evaluation of an advanced local search evolutionary algorithm. In: 2005 IEEE Congress on Evolutionary Computation, vol. 2, pp. 1777–1784. IEEE (2005). http://ieeexplore.ieee.org/xpls/abs_all.jsp?arnumber=1554903
2. Beyer, H.G.: The Theory of Evolution Strategies. Natural Computing Series. Springer, Heideberg (2001)
3. Beyer, H.-G., Sendhoff, B.: Covariance matrix adaptation revisited – the CMSA evolution strategy. In: Rudolph, G., Jansen, T., Lucas, S., Poloni, C., Beume, N. (eds.) PPSN 2008. LNCS, vol. 5199, pp. 123–132. Springer, Heidelberg (2008)
4. Das, S., Suganthan, P.N.: Differential evolution: a survey of the state-of-the-art. IEEE Trans. Evol. Comput. 15(1), 4–31 (2011)
5. Eberhart, R., Kennedy, J.: A new optimizer using particle swarm theory. In: 1995 Proceedings of the Sixth International Symposium on Micro Machine and Human Science, MHS 1995, pp. 39–43, October 1995
6. Fisher, R.A.: The use of multiple measurements in taxonomic problems. Ann. Eugenics 7(2), 179–188 (1936). http://onlinelibrary.wiley.com/doi/10.1111/j.1469-1809.1936.tb02137.x/abstract
7. Gallagher, M.: Clustering problems for more useful benchmarking of optimization algorithms. In: Dick, G., et al. (eds.) SEAL 2014. LNCS, vol. 8886, pp. 131–142. Springer, Heidelberg (2014)
8. Gould, N.I.M., Orban, D., Toint, P.L.: CUTEr and SifDec: a constrained and unconstrained testing environment, revisited. ACM Trans. Math. Softw. 29(4), 373–394 (2003)
9. Hansen, N., Ostermeier, A.: Completely derandomized self-adaptation in evolution strategies. Evol. Comput. 9(2), 159–195 (2001)
10. Hansen, N., Auger, A., Ros, R., Finck, S., Posik, P.: Comparing results of 31 algorithms from the black-box optimization benchmarking BBOB-2009. In: ACM-GECCO Genetic and Evolutionary Computation Conference, pp. 1689–1696, Portland, United States, July 2010. https://hal.archives-ouvertes.fr/hal-00545727
11. Keijzer, M., Merelo, J.J., Romero, G., Schoenauer, M.: Evolving objects: a general purpose evolutionary computation library. In: Collet, P., Fonlupt, C., Hao, J.-K., Lutton, E., Schoenauer, M. (eds.) EA 2001. LNCS, vol. 2310, pp. 231–242. Springer, Heidelberg (2002)
12. du Merle, O., Hansen, P., Jaumard, B., Mladenovic, N.: An interior point algorithm for minimum sum-of-squares clustering. SIAM J. Sci. Comput. 21(4), 1485–1505 (1999). http://epubs.siam.org/doi/abs/10.1137/S1064827597328327

13. Nelder, J.A., Mead, R.: A simplex method for function minimization. Comput. J. **7**(4), 308–313 (1965). http://comjnl.oxfordjournals.org/content/7/4/308
14. Rechenberg, I.: Evolutionstrategie: Optimierung Technischer Systeme nach Prinzipien des Biologischen Evolution. Fromman-Holzboog Verlag, Stuttgart (1973)
15. Ruspini, E.H.: Numerical methods for fuzzy clustering. Inf. Sci. **2**(3), 319–350 (1970). http://www.sciencedirect.com/science/article/pii/S0020025570800561
16. Shen, X., Wong, W.H.: Convergence rate of sieve estimates. Ann. Stat. **22**(2), 580–615 (1994). http://projecteuclid.org/euclid.aos/1176325486
17. Shi, Y., Eberhart, R.: A modified particle swarm optimizer. In: Proceedings of the 1998 IEEE International Conference on Evolutionary Computation, IEEE World Congress on Computational Intelligence, pp. 69–73, May 1998
18. Spaeth, H.: Cluster analysis algorithms for data reduction and classification of objects (1980). http://cds.cern.ch/record/102044
19. Storn, R., Price, K.: Differential evolution a simple and efficient heuristic for global optimization over continuous spaces. J. Global Optim. **11**(4), 341–359 (1997). http://link.springer.com/article/10.1023/A
20. Suganthan, P.N., Hansen, N., Liang, J.J., Deb, K., Chen, Y.P., Auger, A., Tiwari, S.: Problem definitions and evaluation criteria for the CEC 2005 special session on real-parameter optimization. Technical report AND KanGAL Report #2005005, IIT Kanpur, India (2005). http://public.cranfield.ac.uk/sims_staff/wcat/cec2005/sessions/

Distributed Adaptive Metaheuristic Selection: Comparisons of Selection Strategies

Christopher Jankee[1]([✉]), Sébastien Verel[1], Bilel Derbel[2], and Cyril Fonlupt[1]

[1] Université du Littoral Côte d'Opale, LISIC, Calais, France
`jankee@lisic.univ-littoral.fr`
[2] CRIStAL – CNRS – INRIA Lille, Université Lille 1, Lille, France

Abstract. In Distributed Adaptive Metaheuristics Selection (DAMS) methods, each computation node can select, at run-time during the optimization process, one metaheuristic to be executed from a portfolio of available metaheuristics. Within the DAMS framework, we investigate different metaheuristic selection strategies which enable to choose locally at each time step a metaheuristic to execute. We conduct a throughout experimental analysis in order to better understand the accuracy and the behavior of the proposed strategies, as well as their relative performance. In particular, we analyze the impact of sharing metaheuristic performance information between compute nodes and the relative effect on each of the considered distributed selection strategies depending on communication topology. Our experimental analysis is performed on the simple one Max problem, for which the best metaheuristics that should be executed at run-time are known, as well as on the more sophisticated NK-landscapes for which non-linearity can be tuned.

1 Introduction

1.1 Motives

A challenging question accruing in practice when solving an optimization problem using evolution algorithms or metaheuristics is the choice of the relevant algorithm, or at least the choice of the parameters of a given algorithm. This choice should typically be guided by the specific features of the tackled problem, even if in a black-box context, those features could be hard to extract.

In this context, a technique for *algorithm selection* consists in selecting the 'best' algorithm to solve a given problem. The original framework of algorithm selection has been proposed by Rice [12]: First some problem features are extracted. According to those features, one algorithm is selected from a set of available algorithms. Then the performance of the selected algorithm is measured on the problem. With the increasing number of available algorithms, and the number of components that can take part in good algorithms, this framework has become more and more popular. Instead of developing a new optimization algorithm, the "design" of relevant algorithm turns out to the identification of the most suitable one or the most suitable components (See [10] for a recent review on algorithm selection).

© Springer International Publishing Switzerland 2016
S. Bonnevay et al. (Eds.): EA 2015, LNCS 9554, pp. 83–96, 2016.
DOI: 10.1007/978-3-319-31471-6_7

Similarly, the performance of metaheuristics heavily depends on the correct choice of their parameters. Indeed, algorithm selection is related to parameter setting, in the sense that parameters setting can be associated to a specific algorithm, and *vice-versa*. Eiben *et al.* [4] propose to classify parameter setting methods into two classes. In off-line tuning methods, an algorithm is selected before applying it effectively. Some tuning methods use performance prediction methods based on problem features such as in SATzilla [17], and some others are based onsearching in the set of possible algorithm or configurations such as in racing technics [11]. In on-line control methods, the algorithm is selected during the optimization process. At each round, an algorithm is selected from a portfolio of algorithms according to the performance observed in previous rounds. On-line algorithm selection can be modeled as a (dynamic) multi-armed bandit problem: each arm is an optimization algorithm, the reward reflects the quality of solutions produced by the algorithm, and the objective is to select the arms during the optimization process in order to maximize the quality of the final solution. In this context, the so-called Adaptive Operator Selection methods aims at selecting sequentially an operator at each time step. To cite a few, Thierens [14] uses probability matching and adaptive pursuit technics to perform the selection, and Fialho *et al.* [6] propose different selection strategies based on the Upper Confidence Bounds strategy with dynamics restart techniques. For continuous optimization, on-line portfolio techniques have also been recently investigated in [1] using specific reward functions specific to the continuous case.

In this paper, we extend the so-called Distributed Adaptive Metaheuristic Selection (DAMS) framework [3] by investigating on-line portfolio methods in a distributed environment. The DAMS framework is basically motivated by the increasing number of parallel computing facilities (multi-cores, clusters, etc.) and the compute power that can offer when tackling hard optimization problems. DAMS is also tightly related to Evolutionary Algorithms (EAs) based on the Island model [15]. parallelize EAs. In such a model, the population is divided into several subpopulations. Each compute node (an Island) applies an EA on those subpopulations, and the subpopulations can interact within a migration phase where solutions can be exchanged. In the context of on-line portfolio methods, we are interested in a *heterogeneous* island model where each island applies its own and possibly different EA. More precisely, the DAMS framework focuses on setting up adaptive strategies to select at each round a relevant EA which is applied to the local sub-population in order to maximize the performance of the whole distributed system. The goal of this paper is to integrate new distributed adaptive strategies and to study their impact within the DAMS framework. In the rest of this paper, we first review some works related to DAMS. Then, we propose a classification of distributed selection strategies into independent and collective ones according to the information exchange. An experimental analysis is then provided and the impact of the considered strategies is reported.

1.2   Related Work

Two classes of parameters can be controlled in an island model: the parameters related to the migration policy, and the parameters that define the algorithm at each node.

Control of the migration policy: Candan *et al.* [2] propose to control the migration policy on-line in an heterogeneous island model where each island can apply its own EA. A parameter p_{ij} is used to define the migration rate between islands i and j. According to the island performance in producing promising solutions, the rates are updated using a reinforcement learning principle. Ferdandez *et al.* [5] propose a control method of the EA migration policy when the population is 2d-spatially structured following a 2d-grid. The migration, and thus the EA matting, is controlled by moving the solutions on the grid either randomly, or towards a cell surrounded by similar solutions.

Control of the EA parameters: Instead of using the same parameters setting in every island, in a heterogeneous island model, each island applies its own algorithm. In order to demonstrate the usefulness of such heterogeneous model, Tanabe *et al.* [8,13] show that a collection of random parameters provides better performance than a uniform static setting. The study focuses on continuous optimization and differential evolution algorithms, and also on two classes of combinatorial problems (QAP, TSP) using a simple genetic algorithm. Following similar ideas, Garcia-Valdez *et al.* [7] showed that for distributed pool-based EA which is another model of heterogeneous islands, a random set of parameters used by a simple GA on the P-Peaks problems outperforms a static setting.

However, in a heterogeneous island model, random parameter setting is not the only possibility. In fact, each EA associated to each island can be controlled during the optimization process according to state of the search in past iterations. For instance, Tongchim *et al.* [16] proposed to select the parameters (cross-over and mutation) of a simple EA adaptively. Two set of parameters are compared on the same compute node, and the best setting with the best solution is sent to other islands. The authors showed that this kind of on-line mechanism improves over static or random settings.

The DAMS framework [3] proposes to locally select at each round and for each node a metaheuristic from a portfolio of metaheuristics in order to maximize the performance of the whole distributed system. For each compute node using a selection strategy, a metaheuristic is selected not only according to the previous performance of the node, but also according to the performance observed and communicated by neighboring nodes. In their paper [3], Derbel *et al.* propose a simple but yet effective strategy called Select-Best-and-Mutate. In this paper, we propose to analyze other alternative selection strategies taking inspiration from existing multi-arm bandit strategies, but in a distributed (island) model.

2   Adaptive Selection Strategies for DAMS

We first recall the DAMS framework and the original Select-Best-and-Mutate selection strategy. Alternative independent and collective selection strategies based on classical multi-arms bandit strategies are then proposed.

2.1   DAMS and Select-and-Best-Mutate Strategy

The Distributed Adaptive Metaheuristic Selection (DAMS) framework has been introduced in [3]. Algorithm 1 gives the original algorithm using a generic metaheuristic selection strategy. DAMS is a heterogeneous island-like model algorithm. In each compute node, a metaheuristic from a portfolio is applied on the local subpopulation, and the metaheuristic could be different from one node to another. The authors distinguish three basic levels that can be controlled during one round of a DAMS algorithm: the distributed, the metaheuristic selection and the atomic levels. At the distributed level, information between neighboring nodes are shared, migration of solutions is achieved, and the reward of the metaheuristic that has been executed on the node is communicated to neighbors, and *vice-versa*. At the metaheuristics selection level, one metaheuristic is selected from the portfolio according to previously collected rewards. At the last level of the algorithm, called 'the atomic low level' in the original paper, the selected metaheuristic is applied and the corresponding reward is computed.

The authors also proposed the so-called Select-Best-and-Mutate (SBM) to be used at the selection level. SBM strategy is simply based on a metaheuristic mutation rate p_{mut}. With probability $1 - p_{mut}$, SBM selects the metaheuristic having the best reward in the last round among all neighbors (including the current node), and with rate p_{mut}, SBM selects one random metaheuristic from the portfolio $\mathcal{M}$ different from the best one. In others words, SBM has an intensification component that selects the best rewarded metaheuristic at the previous round from the neighboring metaheuristics, and a diversification component that allows to explore new randomly selected metaheuristic. This strategy is related to the well-known ϵ-greedy strategy in multi-armed bandit problem, which selects the arm with the highest estimated expectation with rate $1 - \epsilon$, and uniformly random arm with rate ϵ. In SBM, the reward of metaheuristic is the maximum reward observed in the last round in the node and the neighboring nodes. There is no long-term memory mechanism which computes an estimated average reward from the previous rounds, and the maximum reward is estimated using the neighboring nodes.

2.2   Independent *vs.* Collective Selection Strategies

Similar to the distributed multi-arm bandit problem, in the distributed adaptive portfolio methods, the collaboration of the k compute nodes can contribute to improve the estimation of the quality of metaheuristics, but with an additional communication cost due to information sharing between nodes. Hence, a distributed metaheuristic selection strategy has to take care of this classical trade-offs in distributed systems. Moreover, multi-arm bandit strategies are often a combination of two parts, one exploitation part which promotes the best estimated arm, and one exploration part which looks at new random arms. The exploration part is particularly important when facing a non-stationary problem. The strategy should be able to explore arms for which the reward could have changed. Therefore, when several computation nodes collaborate to improve the metaheuristic

---

Algorithm 1. DAMS algorithm for each computation node

Inputs: A portfolio of metaheuristics $\mathcal{M}$
$r \quad \leftarrow$ INIT_REWARD()
$M \leftarrow$ INIT_META($\mathcal{M}$)
$P \leftarrow$ INIT_POP()
repeat
   | /* Distributed Level:
   | migration and information sharing */
   | Send Msg(r, M, P) to each neighbor
   | $\mathcal{P} \leftarrow \{\} \; ; \mathcal{S} \leftarrow \{\}$
   | **for** *each neighbor w* **do**
   | | Receive Msg(r', M', P') from w
   | | $\mathcal{P} \leftarrow \mathcal{P} \cup \{P'\}$
   | | $\mathcal{S} \leftarrow \mathcal{S} \cup \{(r', M')\}$
   | $P \leftarrow$ UPDATE_POPULATION($P, \mathcal{P}$)
   | /* Metaheuristic Selection Strategy Level */
   | $M \leftarrow$ SELECT_META($\mathcal{M}, (r, M), \mathcal{S}$)
   | /* Atomic Low Level:
   | apply metaheuristic and compute reward */
   | $P_{new} \leftarrow$ APPLY(M, P)
   | $r \quad \leftarrow$ REWARD(P, P_{new})
   | $P \quad \leftarrow P_{new}$
until *Stopping condition is satisfied*;

---

quality estimation, the exploitation part could be reinforced too much forcing the strategy to converge too quickly in a non-stationary scenario.

We distinguish two extreme types of selection strategies according to the information sharing between nodes. In *independent* selection strategies, the metaheuristic selection depends solely on the reward information produced locally by the node. In *collective* selection strategies, the selection takes into account the reward information communicated by the neighboring nodes. For example, the SBM strategy is a collective strategy, and a baseline strategy which selects a metaheuristic uniformly at random is an independent strategy.

2.3 Independent Selection Strategies

First, we can derive a simple independent selection strategy from the original SBM strategy. In fact, instead of selecting the best rewarded metaheuristic from neighboring nodes, we can select the best rewarded metaheuristic in the last W rounds and executed locally by a node – no reward information from neighbors is used. Accordingly, the original collective SBM strategy will be denoted as SBMc, and the newly designed independent SBM by SBMi. Notice that SBMi comes with two parameters, the original mutation rate p_{mut}, and the windows size W.

The so-called Adaptive Pursuit (AP) belongs to the class of probability matching algorithms. AP is a classical adaptive selection strategy used in optimization [14], and can be used as an independent selection strategy. In adaptive pursuit algorithm, a metaheuristic i is applied at time step t in proportion to a probability $p_{i,t}$, and those probabilities are updated according to the rewards of metaheuristics. This technique is then divided into three parts: the update of the reward estimation $\hat{q}_{i,t}$ of the metaheuristics, the update of the probabilities $p_{i,t}$, and the selection of the metaheuristic. Equation 1 defines the update of estimated reward of the metaheuristic i. Variable $r_{i,t}$ is the reward at round t of the metaheuristic i, and parameter $\alpha \in (0,1]$ is the adaptation rate.

$$\hat{q}_{i,t+1} = \hat{q}_{i,t} + \alpha \cdot (r_{i,t} - \hat{q}_{i,t}) \tag{1}$$

The update of the probabilities $p_{i,t}$ is given by Eq. 2 where i_t^* denotes the metaheuristic with the best $\hat{q}_{i,t}$:

$$p_{i,t+1} = \begin{cases} p_{i,t} + \beta \cdot (p_{max} - p_{i,t}), & \text{if } i = i_t^* \\ p_{i,t} + \beta \cdot (p_{min} - p_{i,t}), & \text{otherwise.} \end{cases} \tag{2}$$

For the best estimated metaheuristic, the probability converges to p_{max} with the learning rate β, for the other metaheuristics, the probability converges to p_{min}. At round t, the AP selects the metaheuristic at random in proportion of probability $p_{i,t}$. This independent strategy is denoted by APi.

Several Upper Confidence Bound (UCB) algorithms are used in the context of adaptive metaheuristic selection (see [6] for a review). Let $n_{i,t}$ denotes the number of times the ith metaheuristic is applied up to round t, and let $\hat{q}_{i,t}$ denotes the average empirical reward of metaheuristic i. At each round t, UCB selects the metaheuristic that maximizes the following quantity:

$$\hat{q}_{i,t} + C \cdot \sqrt{\frac{2\log(\sum_j n_{j,t})}{n_{i,t}}}$$

Parameters C enable to control the exploitation / exploration trade-off. This independent selection strategy is denoted by UCBi.

The UCB strategy is an optimal strategy for stationary problems with independent arms which is actually not the case metaheuristics control. The average empirical reward could be far from the current new reward. To overcome this drawback, the average empirical reward can be computed over a slicing windows by considering the last W rounds. This variant is denoted by UCB-Wi.

Finally, a dynamic version of UCB is introduced in [6] and uses the Page-Hinkley test to detect whether the empirical rewards collected for the best metaheuristic have changed significantly. For more details, the reader is referred to page 6 in [6]. This selection strategy will be denoted by UCBP-PHi, and it requires two parameters: a restart threshold γ and a robustness threshold δ.

2.4   Collective Selection Strategies

Each of the above-mentioned independent selection strategies can be used to define a collective selection strategy that takes into account the reward information exchanged with neighboring nodes. In collective SBM which is the original one, the best rewarded metaheuristic is selected from the set of neighboring nodes. In collective AP, the rewards of all neighbors are iteratively used to update the estimation of reward $\hat{q}_i$. Notice that the order of the update could have an impact on the estimation. So, at the initialization phase, a pre-established order between neighboring nodes is randomly chosen. Then, after the updates of reward $\hat{q}_i$, the probability p_i is updated once for all neighbors. In the collective versions of UCB strategies, the empirical average $\hat{r}_i$ is also updated using the rewards of neighboring nodes. The numbers of times $n_{i,t}$ that each metaheuristic is applied is also update according to the information given by each nodes. Notice that in that case, the order of the update does not matter. The selected metaheuristic is the metaheuristic selected after taking into account all neighbors information. Those collective strategies versions are denoted respectively SBMc, APc, UCBc, UCB-Wc, UCB-PHc.

3   Experimental Analysis

3.1   Experimental Setup

Following previous works [2,3,6,16] on adaptive portfolio selection, we also use the well known one-Max problem, which counts the number of 1 in a bit string. In a similar scenario, we use a portfolio of four $(1 + \lambda)$-ES: from one parent solution, the algorithm produces λ solutions according to a stochastic operator and selects the best one for the next iteration. Four operators are used: three operators respectively flip exactly 1, 3 and 5 bits, and the last one uniformly flips each bit with rate $1/N$ where N is the bit strings size set to $N = 1000$.

We use an elitism migration mechanism. Each node (island) sends their current solution to their neighboring nodes. Then, each node receives all solutions from the neighboring nodes. The best solution from the set containing the received solutions and the current solution of the node replace the current solution of the each node. The DAMS algorithm stops when the global maximum is found by one node of the distributed system, when the number of rounds exceeds $T_{limit} = 5.10^4$. 200 runs are computed for each possible strategy and topology. The performance of algorithms is measured either with the number of rounds to reach the global maximum, either using the expected running time (ERT). ERT is expected running time to reach a level fitness of the algorithm with simulated restart. It is equal to $E_s[T] + (1 - \hat{p}_s)/\hat{p}_s.T_{limit}$ where $\hat{p}_s$ the estimated success rate, and $E_s[T]$ is the average number of rounds when the fitness level is reached.

We study four topologies of network: the complete topology where each node is connected to all others nodes, a random topology where there is an edge between two nodes with probability $p = 0.1$, the grid topology which is a two-dimensional regular square grid where each node is connected to the four nearest neighbors, and the circle topology where the nodes are connected to two others nodes to form a circle. The size of the networks is $n \in \{4, 16, 32, 64\}$. In order to have the same number of fitness evaluations in one round whatever the network size n, the λ parameter is set to $64/n$.

A couple of parameters are used in the different selection strategies. For the SBM strategies, the value of metaheuristic mutation rates are $p_{mut} \in$

$\{0.001,\ 0.002, 0.01, 0.1\}$. The window size of the SBMi is set to 5. For AP, the extreme values are set to $p_{min} = 0.1$ and $p_{max} = 1$. The adaptive and the learning rates are $\alpha \in \{0.1, 0.25, 0.5, 0.75, 1\}$ and $\beta \in \{0.1, 0.25, 0.5, 0.75, 1\}$. For all the UCB strategies, the parameter C values are $\{0.1, 0.5, 5, 25, 100\}$. For the variant UCB-W, the set of windows sizes is $\{10, 100, 1000\}$. Following [6], the parameters of Page-Hinkley test are to $\delta = 0.15$, the restart thresholds γ are from $\{0.5, 0.75, 1, 2, 5, 10\}$. Moreover, 2 baseline strategies are used: the random one (rnd.) select at random at each round a metaheuristic, and the constant one (cst.) always select the same metaheuristic which is randomly chosen at the beginning.

3.2   Computational Results

One-Max Overall Performance. From a purely distributed perspective, the first interesting measure is the number of rounds it takes for an algorithm to find the global maximum. The number of rounds provides an idea about the degree of parallelism in an ideal scenario where the communication cost is assumed to be negligible compared to the cost of function evaluation. The relative performance of the different strategies is summarized in Table 1. The best performing parameters are set for each strategies. Several observations can be extracted from Table 1. First, the performance of the different strategies are consistent with the considered configurations in the sense that they can overall be ranked similarly independently of the topology type or graph size. More importantly, we remark that the impact of exchanging rewards information between node has a strong impact on performance. Interestingly, this impact is positive in the case of SBM and AP, whereas it is not when considering UCB. In fact, SBMc appears to overall outperform all the other strategies and APc appears to performing best when both considering the circle, grid and random topologies with large number of nodes. In contrast, the performance of the four implemented versions of UCB is deteriorating systematically as the information from neighbors is incorporated. We attribute this to the fact that this information is actually pushing the UCB strategy to diversify more the search as soon as some operators (even with a good rewards) has been used by other neighboring nodes. UCB is less effective than random selection. The C-value which tunes the exploration-exploitation tradeoff has no impact on this result. Indeed, we have performed an extended sensitive analysis of parameter C (not presented here to save space) which does not changed this result. This also suggests that the UCB strategy has to be completely rethought in order to infer accurate exploration-exploitation tradeoff in the dynamic distributed setting. Notice however, that independent UCB-HPi is still able to provide very competitive results compared to SBMc and APc.

Sensitivity to Parameters. In the previous discussion, we were focused on the overall behavior of the different strategies for a fixed parameter setup. In fact, one may wonder what is the impact is of the parameters used for every strategy. This is illustrated in Fig. 1 where we give representative examples on the sensitivity of SBM, AP and UCB-HP to different parameter settings both in

Table 1. For each topology and graph size, number of selection strategies which statistically outperforms (according to the Wilcoxon test at confidence level $p = 0.05$) a given strategy method for the one-Max problem with $N = 1000$. The 0 value is the best one: no other strategy significantly outperforms the considered one.

Topo.	Size	cst.	rand.	SBMi	SBMc	APi	APc	UCB					
								UCBi	UCBc	HPi	HPc	Wi	Wc
circle	4	8	4	1	0	7	7	10	11	2	3	3	3
circle	16	4	6	3	0	4	0	10	11	1	6	6	6
circle	32	4	6	3	1	4	0	10	11	2	6	6	9
circle	64	4	6	3	2	4	0	10	11	1	6	6	9
grid	4	8	4	1	0	4	9	10	11	2	4	3	3
grid	16	4	5	2	0	4	0	10	11	1	4	6	4
grid	32	4	5	3	1	4	0	10	11	1	4	4	6
grid	64	4	6	3	1	4	0	10	11	1	6	6	9
rnd	4	7	3	0	0	5	7	10	11	0	3	3	3
rnd	16	4	4	1	0	4	3	10	11	1	4	4	5
rnd	32	4	4	3	1	4	0	10	11	2	4	4	9
rnd	64	4	4	3	1	4	0	10	11	1	4	4	9
compl	4	7	3	1	0	7	7	10	10	2	3	3	3
compl	16	6	3	1	0	5	6	11	10	1	4	3	9
compl	32	3	3	2	0	3	8	11	10	1	3	3	9
compl	64	3	3	2	0	3	3	11	10	1	3	7	9
Average		4.875	4.312	2	0.4375	4.375	3.125	10.187	10.75	1.25	4.187	4.437	6.562

the case of an independent and a collective strategy. We can appreciate that SBM is rather stable under different configurations although for the collective variant, the impact of the mutation rate is slightly more pronounced (a small values is advised). The same thing holds for the AP strategy where the algorithm is robust to a wide range of values of α and β, with he exception of the adaption rate $\alpha = 1$ which is to be avoided since it promotes strong convergence in the reward estimation. For the UCB-HP strategy, the value of C, which appears in the confidence bound, plays an important role but only in the independent strategy. For the collective strategy, where the information from neighbors is actually deteriorating performance, the C-value does not seem to have any impact and cannot help obtaining improved results.

Parallelism. In the previous discussions, we were only interested in analyzing the relative behavior of the strategies for a fixed topology. In particular, the results of Table 1 do not allow us to appreciate the relative impact of different topologies on the performance of each strategy. For this purpose, we show in Fig. 2 the relative performance of SBM and AP in different configurations. It is important to recall that the number of function evaluations at every single round and for all the considered configurations is the same which means that the number of function evaluations needed overall in any of the considered configuration

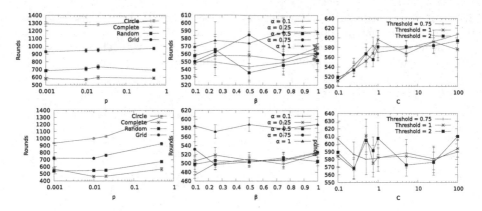

Fig. 1. Average number of rounds to find the maximum of the one-Max problem as function of the parameter values of different selection strategies. From left to right: SBM, AP, UCP-HP strategies ; top: independent selection, bottom collective selection.

is by the *same* multiplicative factor similar to the number of rounds depicted in Fig. 2. This observation has an important impact, since then, we are able to obtain different trade-offs when considering the number of exchanged messages as an important indicator of *parallel speed-ups* that one could obtain when effectively deploying our strategies in a real distributed setting. In fact, the number of messages needed to exchange information is exactly the number of rounds times the number of edges used in the considered topology. In the case of the complete (resp, circle, grid, random) topology, the number of edges is $n(n-1)/2$ (resp. $n-1$, $O(n)$, $O(p.n^2)$) where n is the number of nodes. From Fig. 2, we can notice that the number of rounds stays stable for the complete and random topology (except for 4 nodes) with the complete topology being slightly better. However the number of rounds increases sharply for the circle and the grid which we attribute to the increase of the topology diameter. Roughly speaking, although the increase in the number of rounds for the circle and the grid is at most by a factor of 2, the number of needed messages stays linear in the number of nodes. This is to contrast with the complete topology where the increase in the number of messages is polynomial. Hence, in a practical setting where the cost of message-passing is non-negligible, we claim that the best choice would be the random topology which exhibits the most appealing tradeoffs in terms of the number of rounds *v.s.* the number of messages exchanged overall.

NK-Landscapes. In this paper, we also consider a more sophisticated class of problems captured by the so-called NK-landscapes. The family of NK-landscapes constitutes a model of multimodal problems [9]. The search space is binary strings of size N: $\{0,1\}^N$. N refers to the problem size, and K to the number of bits that influence a particular position from the bit-string, *i.e.* the epistatic interactions. The objective function $f : \{0,1\}^N \to [0,1)$ to be maximized is defined as follows.

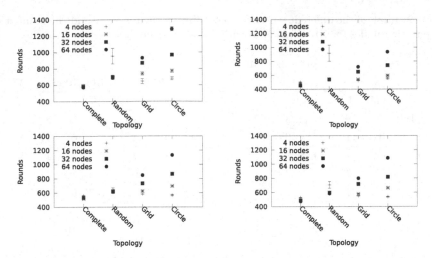

Fig. 2. Average number of rounds to find the maximum (one-Max problem) according to the topology and the number of nodes. From left to right and top to bottom: SBMi, SBMc, APi, and APc strategies.

$$f(x) = \frac{1}{N} \sum_{i=1}^{N} f_i(x_i, x_{i_1}, \ldots, x_{i_K})$$

where $f_i : \{0,1\}^{K+1} \to [0,1)$ defines the component function associated with each bit x_i. By increasing the number of epistatic interactions K from 0 to $(N-1)$, NK-landscapes can be gradually tuned from smooth to rugged. In this work, we set the position of these interactions at random. Component values are uniformly distributed in the range $[0,1)$.

Our interest in the NK-landscapes stems from the fact that usually different bit-flip mutation rates are believed to provide different performances. To illustrate this claim, we show in Fig. 3, the empirical probability that a solution with the fitness given by the x-axis is be improved if a uniform bit-flip operator with rate c/N is applied, where c varies in the range $\{1, 2, 4, 8, 16\}$. We can clearly see that the operator which is likely to provide an improvement depends strongly of the attained fitness level. Hence, this kind of landscapes appears to be particularly interested to be studied within the DAMS framework. Accordingly, we perform the same experiments while considering different NK-landscapes with $N = 1000$ and $K \in \{1, 4, 8\}$. The portfolio of metaheuristics is composed by five $(1 + \lambda)$-ES based on the uniform bit-flip rate c/N with rates $c = 1, 2, 4, 8$, and 16. We tune the parameters according to the results the one-Max problem: $p_{nut} = 0.01$ for SMB, $\alpha = 0.5$ and $\beta = 0.5$ for AP, and $C = 25$ for UCB strategies. Interestingly, we find that no significant differences can be reported between any of the considered selection strategies when looking at the final fitness value (this is nor reported due to space limitations). However, we are able to report different behavior when examining the empirical expected running time (ERT) to attain the median fitness value (computed over all configurations).

Table 2. Rank of the different strategies according to the topology and the number of computation nodes for NK-landscapes with $N = 1000$ and $K = 1, 4, 8$.

Topo.	Size	unif.	cst.	rand.	SBMi	SBMc	APi	APc	UCB					
									UCBi	UCBc	HPi	HPc	Wi	Wc
		$K = 1$												
compl	16	0	9	6	3	7	12	2	1	10	4	5	8	11
compl	64	0	10	7	6	1	4	9	12	3	2	5	11	8
circle	16	0	2	11	3	5	1	10	8	7	6	4	12	9
circle	64	0	7	8	2	1	6	4	12	9	3	10	11	5
average		0	7	8	3.5	3.5	5.75	6.25	8.25	7.25	3.75	6	10.5	8.5
		$K = 4$												
compl	16	0	6	12	9	1	11	3	2	4	8	10	5	7
compl	64	0	6	3	8	5	11	12	1	4	10	2	7	9
circle	16	1	11	12	8	4	5	6	3	7	0	2	10	9
circle	64	0	10	9	11	6	7	12	5	4	3	2	1	8
average		0.25	8.25	9	9	4	8.5	8.25	2.75	4.75	5.25	4	5.75	8.25
		$K = 8$												
compl	16	1	3	9	0	11	7	6	2	10	4	8	5	12
compl	64	0	12	4	10	3	6	9	11	2	5	8	1	7
circle	16	7	0	4	5	6	12	3	9	10	2	1	8	11
circle	64	0	2	12	3	11	8	9	5	10	7	1	4	6
average		2	4.25	7.25	4.5	7.75	8.25	6.75	6.75	8	4.5	4.5	4.5	9

The ERT results are summarized in Table 2. In addition to adaptive selection strategy, we also tested a uniform and static strategy, denoted unif in the table, where every nodes share the same metaheuristic all along the execution. In the table, we choose to present the performance of the best uniform-static strategy which is not the same according to the topology and the number of nodes. Perhaps, the most interesting observation is than the uniform-static strategy is the best performing and none of the considered DAMS variants is able to outperform it. This might be surprising at first sight, but not if we account for the time required to learn the best metaheuristic to apply. In fact, when examining carefully Fig. 3 in light of the information given by the empirical improvement probability, we can see that the fitness level is increasing very abruptly for NK-landscapes in the early stages of the search. Hence, the different fitness windows where one has to choose the best operator are very tight which is to contrast with the time it may need for a strategy to detect which operator is actually the best to apply. As a consequence, even though the fixed operator used by a uniform static strategy is not optimal in all the stages of the execution, it still does not loose time in learning by testing less efficient operators. It worth-noticing that the previous experiments raise the question of whether we really need to adapt the search heuristics at runtime and does it really serve in practice? We argue that

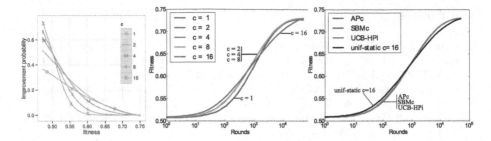

Fig. 3. Empirical improvement probabilities *vs.* fitness level (left). Fitness *vs.* rounds in log-scale. Center: uniform-static, right: different strategies. NK-landscapes with $K = 4$.

the answer to this question is definitively yes. In fact, the general lessons that we can learn from our experiments with the NK-landscapes can be formulated as following. First, in a black-box scenario, the time during which a metaheuristic is the best one depends strongly on the landscape. Hence, learning this landscape at runtime is for sure a plausible alternative. Second, we need to study more carefully the cost of the learning stage of selection strategy in function of the considered landscape, and to design novel alternative adaptive strategy that would be able to minimize the learning cost at the aim of improving efficiency.

4   Conclusion

In this paper, we investigate new adaptive strategies for distributed metaheuristic selection. Accordingly, we explored the applicability of adaptive pursuit and upper bound confidence based algorithms in the distributed setting where several heterogeneous islands have to cooperate in order to select the most accurate metaheuristic dynamically at runtime. In particular, we consider the possibility of incorporating the distributed information coming from the neighboring islands and study its impact on the search behavior by considering independent and collective schemes. We conduct a throughout experimental study in order to better understand the major ingredients toward making such schemes successful. We find that special care must be taken when attempting to use the rewards observed distributively at different islands in order to obtain accurate exploration-exploitation trade-offs. Besides, our study keeps open several questions that deserve further investigation in the future. For instance, we could analyze the selection strategies on others benchmarks such as knapsack or graph coloring problems. It would also be interesting to study the gain one can achieve by the proposed strategies when effectively deployed in a real distributed testbed. In such a setting, the communication cost is very likely to introduce new challenges; but the increasing power offered by modern computation systems is worth to be investigated in order to derive highly efficient adaptive strategies.

References

1. Baudiš, P., Pošík, P.: Online black-box algorithm portfolios for continuous optimization. In: Bartz-Beielstein, T., Branke, J., Filipič, B., Smith, J. (eds.) PPSN 2014. LNCS, vol. 8672, pp. 40–49. Springer, Heidelberg (2014)
2. Candan, C., Goeffon, A., Lardeux, F., Saubion, F.: A dynamic island model for adaptive operator selection. In: GECCO 2012, pp. 1253–1260 (2012)
3. Derbel, B., Verel, S.: DAMS: distributed adaptive metaheuristic selection. In: GECCO 2011, pp. 1955–1962 (2011)
4. Eiben, A.E., Michalewicz, Z., Schoenauer, M., Smith, J.E.: Parameter control in evolutionary algorithms. In: Lobo, F.G., Lima, C.F., Michalewicz, Z. (eds.) Parameter Setting in Evolutionary Algorithms. Studies in Computational Intelligence, vol. 54, pp. 19–46. Springer, Heidelberg (2007)
5. Fernandes, C.M., Laredo, J.L.J., Merelo, J.J., Cotta, C., Nogueras, R., Rosa, A.C.: Shuffle and mate: a dynamic model for spatially structured evolutionary algorithms. In: Bartz-Beielstein, T., Branke, J., Filipič, B., Smith, J. (eds.) PPSN 2014. LNCS, vol. 8672, pp. 50–59. Springer, Heidelberg (2014)
6. Fialho, A., Da Costa, L., Schoenauer, M., Sebag, M.: Analyzing bandit-based adaptive operator selection mechanisms. AMAI **60**, 25–64 (2010)
7. García-Valdez, M., Trujillo, L., Merelo-Guérvos, J.J., Fernández-de-Vega, F.: Randomized parameter settings for heterogeneous workers in a pool-based evolutionary algorithm. In: Bartz-Beielstein, T., Branke, J., Filipič, B., Smith, J. (eds.) PPSN 2014. LNCS, vol. 8672, pp. 702–710. Springer, Heidelberg (2014)
8. Gong, Y., Fukunaga, A.: Distributed island-model genetic algorithms using heterogeneous parameter settings. CEC **2011**, 820–827 (2011)
9. Kauffman, S.A.: The Origins of Order. Oxford University Press, Oxford (1993)
10. Kotthoff, L.: Algorithm selection for combinatorial search problems: a survey. AI Mag. **35**, 48–60 (2012)
11. López-Ibáñez, M., Dubois-Lacoste, J., Stützle, T., Birattari, M.: The R package irace package, iterated race for automatic algorithm configuration. Technical report TR/IRIDIA/-004, IRIDIA, (2011)
12. Rice, J.R.: The algorithm selection problem. Adv. Comput. **15**, 65–118 (1976)
13. Tanabe, R., Fukunaga, A.: Evaluation of a randomized parameter setting strategy for island-model evolutionary algorithms. In: CEC 2013, pp. 1263–1270 (2013)
14. Thierens, D.: An adaptive pursuit strategy for allocating operator probabilities. In: GECCO 2005, pp. 1539–1546 (2005)
15. Tomassini, M.: Spatially Structured Evolutionary Algorithms: Artificial Evolution in Space and Time. Natural Computing Series. Springer, Heidelberg (2005)
16. Tongchim, S., Chongstitvatana, P.: Parallel genetic algorithm with parameter adaptation. Inf. Process. Lett. **82**(1), 47–54 (2002)
17. Xu, L., Hutter, F., Hoos, H.H., Leyton-Brown, K.: Satzilla: portfolio-based algorithm selection for sat. J. Artif. Int. Res. **32**(1), 565–606 (2008)

Combining Mutation and Recombination to Improve a Distributed Model of Adaptive Operator Selection

Jorge A. Soria-Alcaraz[1]([✉]), Gabriela Ochoa[2], Adrien Göeffon[3],
Frédéric Lardeux[3], and Frédéric Saubion[3]

[1] Depto de Estudios Organizacionales, Universidad de Guanajuato-División
de Ciencias Económico-Administrativas, Guanajuato, Mexico
jorge.soria@ugto.mx
[2] University of Stirling, Stirling, UK
[3] University of Angers, Angers, France

Abstract. We present evidence indicating that adding a crossover island greatly improves the performance of a Dynamic Island Model for Adaptive Operator Selection. Two combinatorial optimisation problems are considered: the Onemax benchmark, to prove the concept; and a real-world formulation of the course timetabling problem to test practical relevance. Crossover is added to the recently proposed dynamic island adaptive model for operator selection which considered mutation only. When comparing the models with and without a recombination, we found that having a crossover island significantly improves the performance. Our experiments also provide compelling evidence of the dynamic role of crossover during search: it is a useful operator across the whole search process. The idea of combining different type of operators in a distributed adaptive search model is worth further investigation.

1 Introduction

Search operators are key elements of heuristic search algorithms, determining the structure of the fitness landscape being searched. A large variety of operators have been proposed in the literature for combinatorial optimisation problems. However, given a new problem or instance of a combinatorial problem it is not clear before hand which operator (or indeed set of operators) will be the most effective. In response to this, modern heuristic approaches combine several operators. Some schemes such as variable neighbourhood search, or standard memetic algorithms combine operators in a pre-determined way. Some other schemes, such as hyper-heuristics [2,12], and adaptive operator selection approaches [10], acknowledge the advantage of combining a pool of operators; but most importantly, they also realise that the usefulness of specific operators can vary dynamically across the search process. Therefore, they propose adaptive, learning-based mechanisms for selecting operators on the fly.

Island models [18] were initially introduced for avoiding premature convergence in evolutionary algorithms (EAs). They use a set of sub-populations

© Springer International Publishing Switzerland 2016
S. Bonnevay et al. (Eds.): EA 2015, LNCS 9554, pp. 97–108, 2016.
DOI: 10.1007/978-3-319-31471-6_8

instead of a single a panmictic one. Sub-populations evolve independently on separated islands during some search steps and interact periodically with other islands by means of migrations [14], whose impact has been carefully studied [8,9]. Two main types of island models can be considered. First, replicating the same algorithm on each island with the view of improving the management of the population. This constitutes the most common use of island models and is closely related to distributed evolutionary algorithms [9]. Second, considering different algorithms (or algorithms settings) on each island as a dynamic control method in order to identify the most promising algorithm according to the current state of the search.

Island models traditionally use fixed migration policies in order to reinforce the islands characteristics [1,6,15]. An alternative dynamic migration policy was proposed by Lardeux and Goëffon [7], where migration probabilities change during the evolutionary process according to the impact of previous analogue migrations. The island model should be able to both identify the current most appropriate subset of islands for improving individuals, and to quickly react to changes if other heuristics (operators) turn out to be more beneficial.

It is important to stress that in this article, the island model does not implement a complete evolutionary algorithm in each island as it is usually done. Instead, each island is associated with a single (different) search operator, and in every iteration the island's operator is applied to all individuals in the island. This constitutes an approach to adaptive operator selection as recently proposed by Candan et al. [3]. The ability of the dynamic island model to efficiently manage simple operators has already been compared to other adaptive operator selection approaches in [3]. So far, mutation operators or abstract scenarios have been considered. The motivation of this paper is to assess the efficiency of the island model in presence of different kinds of operators, such as crossover on various problems. The idea is to assign an operator to each island and use the dynamic regulation of migrations to distribute the individuals on the most promising islands (i.e., the most efficient operators) at each stage of the search.

The main contribution of this article is the introduction of crossover in conjunction with mutation operators, while the original adaptive operator selection model considered only mutation operators [3]. In our proposal, individuals from different islands can undergo recombination when they "visit" the recombination island and thus may directly share information. We found that having a crossover island significantly improves the model's performance. We demonstrate this by comparing the models with and without recombination on two selected benchmarks: the Onemax problem, widely used to prove concepts in adaptive operator selection studies [3–5]; and a formulation of the course timetabling problem considering the set of publicly available real-world instances from the 2007 International Timetabling Competition *ITC-2007* [11].

The article is organised as follows. Section 2 introduces the dynamic island model of adaptive operator selection, and how we incorporated crossover into it. Section 3 describes the experimental setting, while results are presented in Sect. 4. Finally, Sect. 5 summarises our findings and suggests directions for future work.

2 Crossover as an Island Operator

We start by formally presenting the dynamic island model for adaptive operator selection and follow by describing how crossover was incorporated.

2.1 Dynamic Island Model

Let us consider an optimisation (minimisation) problem defined as a pair $(\mathcal{S}, f)$ where $\mathcal{S}$ is a search space whose elements represent candidate solutions of the problem, and $f : \mathcal{S} \rightarrow \mathbb{R}$ is an objective function. An optimal solution is an element $s^* \in \mathcal{S}$ such that $\forall s \in \mathcal{S}, f(s^*) \geqslant f(s)$.

An Island Model can be formally defined as a tuple $(\mathcal{I}, \mathcal{H}, \mathcal{P}, V, M)$. Where $\mathcal{I} = \{i_1, \cdots, i_n\}$ is the set of Islands, $\mathcal{H} = \{H_1, \cdots, H_n\}$, a set of heuristics (operators in this paper), and $\mathcal{P} = \{p_1, \cdots, p_n\}$ a collection of sub-populations, one per island. The topology of the model is given by an undirected graph $G(\mathcal{I}, V)$ where $V \subseteq \mathcal{I}^2$ is a set of edges between islands ($\mathcal{I}$, the nodes of the graph.) Finally, the migration policy is given by a square matrix M of size n, such that $M(i,j) \in [0,1]$ represents the probability for an individual to migrate from island i to island j. Each island k is equipped with a sub-population p_k and an operator H_k. The matrix M is coherent with the topology, i.e., if $(i,j) \notin V$ then $M(i,j) = 0$. Algorithm 1 outlines the operation of an Island Model for minimisation problems.

Algorithm 1. Basic Island Model

Require: an IM $(\mathcal{I}, \mathcal{A}, \mathcal{P}, V, M)$, an Optimisation problem $(\mathcal{S}, f)$
 1: **while** not stop condition **do**
 2:   **for** $i \leftarrow 1$ **to** n **do**
 3:     $p_i \leftarrow H_i(p_i)$
 4:     **for** $s \in p_i$ **do**
 5:       **for** $j \leftarrow 1$ **to** n **do**
 6:         generate a random number $rand$
 7:         **if** $rand < M(i,j)$ **and** $|p_i| > 0$ **then**
 8:           $p_j \leftarrow p_j \cup \{s\}$
 9:           $p_i \leftarrow p_i \setminus \{s\}$
10:         **end if**
11:       **end for**
12:     **end for**
13:   **end for**
14:   $b \leftarrow best(\bigcup_i (p_i))$
15:   **if** $f(b) > f(s^*)$ **then**
16:     $s^* \leftarrow b$
17:   **end if**
18: **end while**
19: **return** s^*

In the algorithm, p_i denotes the sub-population at island i and $H_i(p_i)$ (line 3) the population obtained after applying heuristics H_i on it. The function $best$

computes the best current individual w.r.t. objective function f. The stopping condition is, as usual, a limited number of iterations or the fact that an optimal solution has been found in the global population. The migration matrix M is used to send individuals to other islands or stay on the same one.

In dynamic island models, an adaptive update of the migration matrix at iteration $t + 1$, denoted M_{t+1}, is performed as:

$$M_{t+1}(i, k) = (1 - \beta)(\alpha.M_t(i, k) + (1 - \alpha)R_{i,t}(k)) + \beta N_t(k)$$

where N_t is a stochastic noise vector such that $||N_t|| = 1$ and $R_{i,t}$ is a reward vector that is computed after applying H_i at time t. α allows to control the balance between previous knowledge accumulated and immediate observed effect. β controls the amount of noise, which is necessary to explore alternative actions. These parameters need to be tuned and their impact has been studied in [3]. The reward $R_{i,t}(k)$ is defined as:

$$R_{i,t}(k) = \begin{cases} \frac{1}{|B|} & \text{if } k \in B, \\ 0 & \text{otherwise,} \end{cases}$$

where B is the set of the operators that have been produce the best improvement for each island i.e., operators producing the best improvements according to f for each island at a given time.

2.2   Incorporating Crossover

Mutation heuristics perform a change on a given solution, by swapping, changing, removing, adding or deleting solution components. In contrast crossover operators, take two (or more solutions), combine them and return a new solution (or more than one solution).

Let $s \in S$ be a solution. A (unary) mutation operator can be formally defined as $H_m : S \rightarrow S$. Crossover operators can in turn, be defined with the following signature $H_c : S \times S \rightarrow S \times S$. We propose to incorporate crossover as an island operator. The key idea is to define the crossover H_c with a similar formal signature than mutation H_m.

Algorithm 2. Standard Operator Island

Require: a population p
1: OffspringPool $= \emptyset$
2: **for all** $s \in p$ **do**
3:     OffspringPool $=$ OffspringPool $\cup \{H(s)\}$
4: **end for**
5: **return** OffspringPool

Algorithm 2 outlines the behaviour of an operator island in the island model. The operator H is applied at line **3**. The crossover island uses the same overall

Algorithm 2, but to apply recombination (H_c) with the same signature than mutation, it requires a single solution as parameter. The crossover is performed using the incoming solution as one parent. The other parent is either a random solution (only for the first iteration) or the last incoming solution. The best generated offspring is then returned. This is outlined in Algorithm 3. With this simple mechanism we can combine mutation and recombination operators in the island model for adaptive operator selection.

Algorithm 3. Crossover Operator H_c

Require: s incoming solution
 1: **if** $Temp$ is undefined **then**
 2:     $Temp = randomSolution()$
 3: **end if**
 4: $Offsprings = Crossover(Temp, s)$
 5: $Temp = s$
 6: **return** $Best(Offsprings)$

---

3   Experimental Setup

Two algorithm variants are considered: *DIM-M*, a dynamic island model of adaptive operator selection with mutation operators only, and *DIM-MX*, which combines mutation and recombination. They are tested using the benchmark problems and algorithm setting described below.

Onemax: (or counting ones problem), is a unimodal maximisation problem traditionally used in theoretical and proof of concept studies in genetic algorithms, where the string of all ones is the single optimum. Following Candan et al. [3] we use a Onemax instance of size $n = 1000$, the algorithm parameters are summarized in Table 1. Four mutation operators and one recombination operator are considered. Each operator is assigned to an island and it is applied regardless of whether it improves or not the incoming solution. The operators used are:

- *bit-flip Mutation:* flips each bit with probability $1/n$.
- *k-bit Mutation:* (with $k = 1, 3, 5$), chooses uniformly at random k bits in the current solution and flips their values.
- *1-point Crossover:* chooses uniformly at random position in the string, and interchanges the sub-strings to produce offspring.

Course Timetabling: is a minimisation problem where the objective is to assign several events to time-slots without violating certain constraints. The problem can be defined in terms of a set of events (courses or subjects) $E = \{e_1, e_2, \ldots, e_n\}$, a set of time-periods $T = \{t_1, t_2, \ldots, t_s\}$, a set of places (classrooms) $P = \{p_1, p_2, \ldots, p_m\}$, and a set of agents (students registered in the courses) $A = \{a_1, a_2, \ldots, a_q\}$. An assignment is then given by the quadruple $(e \in E, t \in T, p \in P, a \in A)$, and a solution to the problem is a complete set of n assignments (one for each event) that satisfies the set of hard constraints. Our formulation uses a generic modelling approach where solutions are encoded as vectors of integer numbers of length equal to the number of events (courses) [16,17]. Positions in the vector represent events, and their integer values are indices in a set of data structures encoding pairs of valid time-slots and class-rooms for each event [16]. A set of four mutation operators are considered, which were the best performing in [17]. They range from simple randomised exchange or swap neighbourhoods to greedy and more informed procedures. As a crossover operator we implemented the simple *1-point* crossover. This is possible with the representation used (a vector if integer numbers) where offspring generated by *1-point* crossover are valid solutions.

- *Simple Random Perturbation (SRP):* uniformly at random chooses a variable and changes its value for another one inside its feasible domain.
- *Swap (SWP):* selects two variables uniformly at random and interchanges their values.
- *Statistical Dynamic Perturbation (SDP):* chooses a variable following a probability distribution based on the frequency of variable selection in the last k iterations. Variables with lower frequency will have a higher probability of being selected. Once selected, the value is randomly changed.
- *Double Dynamic Perturbation (DDP):* similar in operation to SDP, but internally maintains an additional solution, and returns the best of the two solutions.
- *1-point Crossover*: chooses uniformly at random position in the vector, and interchanges the sub-portions to produce offspring.

The experiments considered the 24 real-world instances from the 2007 International Timetabling Competition *ITC-2007*, track 2, which correspond to the post-enrollment course timetabling benchmark[1]. These instances range from 400 to 600 events. Table 1 reports the algorithm parameters used. Experiments were conducted on a CPU with Intel i7, 8 GB Ram using the Java language and the 64 bits JVM.

4   Results

4.1   Onemax

Figure 1 illustrates an example run of the two algorithm variants on the Onemax problem. *DIM-M* contains 4 islands, one for each mutation operator, while *DIM-MX* has 5 islands, corresponding to the 4 mutations and the *1-point* crossover. The curves show, for each operator (island), the sub-population size over time measured as iterations, and reported at intervals of length 150 (the X values are $\times 10$). We consider an iteration as a single complete execution of the DIM algorithm, which this corresponds to a move or migration of individuals across

---

[1] Available at http://www.cs.qub.ac.uk/itc2007/.

Table 1. Algorithm parameters for the two benchmark problems.

Parameter	Onemax	Course timetabling
Chromosome length	1000	400 to 600
Population size	800	1000
Number of islands	4 or 5 (one for each operator)	
Initial migration	1/ number-of-islands	
(α, β)	(0.8,0.1)	(0.8,0.1)
No. of runs	10	10 per instance
Stop criteria	Optimum is found	540 s

islands. The plot also shows (the black solid line) the best individual fitness over time, with values visible in the right-hand axis. The variant without crossover (*DIM-M*, left plot) required over two minutes (128.32 s) to reach the global optimum, which corresponds to nearly 7,000 iterations and 68,251 functions calls. The plot shows how the most explorative *5-bit* operator has the highest attraction rate at the very early stages of the search. Soon, after 50 iterations or so, this rate goes down leaving a less perturbative operator (namely, *1-bit*) to take the lead in the search process. The variant with recombination (*DIM-MX*, right plot) reached the optimum much faster, in less than 30 s, which corresponds to 1,300 iterations and 17,363 functions calls. The plot illustrates the run up to 7,000 iterations for comparisons purposes with the *DIM-MX* variant. In this case, the crossover operator attraction rate increases steadily up to the point where the optima solution is found. Another interesting observation from these experiments is the superiority of the *1-bit* mutation over the more standard *bit-flip* operator for this problem.

Figure 2 offers a close up of the first 3,000 iterations showing population size at each step and considering only two operators for each variant: *1-bit* and *5-bit* for *DIM-M* and, *1-bit* and *crossover* for *DIM-MX*. Note that the horizontal axis shows multiples of 10 iterations. As the right plot of Fig. 2 illustrates, crossover is increasingly useful for *DIM-MX* search up to the point where the optimal solution is found, which occurs around iteration 1,300. This confirms an interesting property of crossover, which was observed by Ochoa et al. before [13]. Crossover is a versatile operator, its role is dynamic: when there is high diversity in the population such as at the beginning of the search process, it acts as an explorative operator. However, when the population diversity is low (i.e., the population is largely converged) it acts instead an improvement operator preserving the useful building blocks. For Onemax, it is clear that, at the beginning of the search when individuals have low quality (i.e., contain few ones) and are very different, crossover may quickly generate new individuals with more ones by recombination and thus quickly explore more interesting areas. While when the population has converged to higher quality (i.e., when individuals contains mainly ones), crossover may also be useful by preserving the components of the highly fit individuals. The probability of selecting crossover eventually

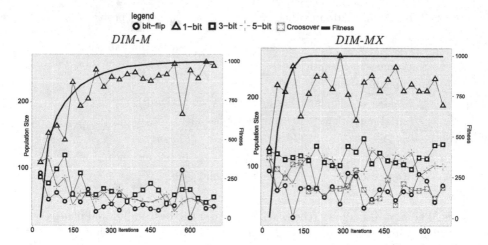

Fig. 1. Onemax. Attraction rate (sub-population size) of each operator (island) along with best fitness over time. Values in the X axis multiplied by 10 give iterations. *DIM-M*, using mutation operators only. *DIM-MX*, combining mutation with a crossover operator.

drops after the optimal solution is found (iteration 1,300) and the performance curve flattens. This is probably due to the computational overhead of crossover as compared to mutation operators. So, it ceases to be selected when no additional improvements are found in the search process. But clearly the operator was increasingly useful from the early stages of the search up to the point when the optimal solution was found. Therefore, crossover is a useful operator across the whole search process.

This contrasts with the behaviour of *5-bit* on the left plot of Fig. 2 (*DIM-M*), where *5-bit* acts an efficient explorative operator early on (up to iteration 500 or so), but then it stops being useful, as it becomes too disruptive and its rate drops (which has also been observed in [3]).

4.2   Course Timetabling

As a first experiment, we ran the two algorithm variants for two minutes (120 s) on a selected course timetabling instance. Specifically, instance number 1 from the *ITC-2007* track 2 set, which consists of 500 students, 400 courses, 35 timeslots and 10 classrooms. Again, *DIM-M* contains 4 islands, one for each mutation operator, while *DIM-MX* has 5 islands, corresponding to the 4 mutations and the *1-point* crossover. Figure 3 illustrates the results. The curves show, for each operator, the sub-population size over time (measured as iterations, at intervals of length 250). The black solid line in the plots shows the best individual fitness over time, with values indicated in the right-hand axis. In this case, we are dealing with a minimisation problem. It can be seen that the number of iterations is 9250 for *DIM-M* (left plot), while it is of 6800 iterations for *DIM-MX*. This is because an *DIM-MX* iteration uses more resources as it consists of 5 operators. Despite

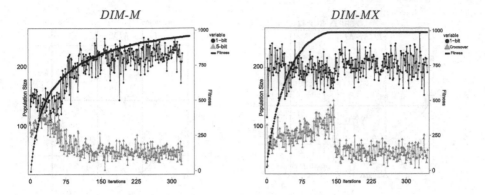

Fig. 2. Onemax. Close up of the attraction rate (sub-population size) of each operator (island) along with best fitness over time, for the first 3,000 iterations. Values in the X axis multiplied by 10 give iterations. *DIM-M*, illustrating *1-bit* and *5-bit*. *DIM-MX*, illustrating *1-bit* and *crossover*

this increased CPU demand, the variant with crossover produces the best results at the end of the 120 s run. Specifically, *DIM-MX* finds a solution with fitness 582 (as seen in the right axis with fitness values), which is a much better value (we're mimimising soft-constraints violations) than the 845 solution achieved by *DIM-M*. The dynamic rates of the operators across the run is more complex for this problem than for the Onemax (Figs. 1 and 2). The operators combine efforts and take turns in solving the problem. The curves, however, indicate that when recombination is not used (*DIM-M*, left plot), the swap (SWP) operator dominates the search, specially at the initial and middle stages, while for the *DIM-MX* variant (right plot), crossover dominates at several stages and enhances the search process.

For a more thorough comparison, we used the experimental conditions and rules followed in the timetabling competition. Specifically, we used the benchmark program provided in the competition site to measure the allowed running time on a given machine. This time is generally between 300 and 600 s (per run, per instance) on a modern PC. Following the competition protocol, 10 replicas per instance were considered, and the remaining algorithm parameters are reported in Table 1.

Table 2 shows results over some representative instances. The variant with recombination *DIM-MX*, consistently produced the best results across all the instances. Moreover, results with *DIM-MX* show a much lower standard deviation. We suggest that this occurs because crossover guides the search by combining information from the whole population, and contributes to escape local optima. For the mutation-only variant, migration among islands is the only mechanism for information exchange. It is more likely in this case for an island to be trapped in a local optima.

A statistical analysis of the results across all test instances was also conducted. Normality and Homocesticity of the data was checked using Shapiro-wilk test. The results of a two-way ANOVA test combining the 24 test instances and

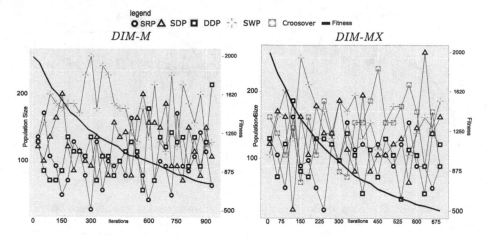

Fig. 3. Course timetabling, instance *ITC-2007-1*. Attraction (sub-population size) of each operator (island) along with best fitness over time. Values in the X axis multiplied by 10 give iterations. *DIM-M*, using mutation operators only. *DIM-MX*, combining mutation with a crossover operator.

Table 2. Course timetabling. Representative *ITC-2007* instances. Results are shown in the form of: $\bar{X}_\sigma$

Instance no.	1	4	10	15	18	20	23
DIM-M	$345.22_{45.23}$	$690.56_{62.49}$	$2778.2_{210.4}$	$30.4_{12.1}$	$40.15_{32.84}$	$186.14_{38.12}$	$1677.14_{420.2}$
DIM-MX	$131.16_{40.10}$	$586.31_{37.78}$	$2358.2_{165.3}$	$7.7_{5.3}$	$22.16_{22.30}$	$150.10_{15.2}$	$1378.4_{290.3}$

Table 3. Course timetabling. Two-way ANOVA F test, pairwise t test and Tukey HSD test.

ANOVA	Df	Sum sq	Mean sq	F value	Pr($>$F)
Algorithm	1	736576	736576	54.771	**6.59e-13**
Instance	23	1699991740	7390945	549.58	$<$2.2e-16
Residuals	455	6118956	13448		
TukeyHSD	diff	lwr	upr	**p adj**	
DIM-5 **vs** *DIM-4*	-78.34	-99.15	-57.54	**0.00**	

2 algorithm variants is reported in Table 3. The test indicates whether (or not) the means of several groups are equal, which in this context refers to whether the competing algorithms have the same performance across the tests instances. The obtained results support the existence of significant performance differences between the DIM variants.

The numbers in bold font under the $(Pr(> F))$ label in Table 3 show the corrected *p-value*. This value represents the probability of obtaining a test statistic result at least as extreme or as close to the one that was actually observed, assuming that the null hypothesis is true (H_0 : algorithms have the same performance).

Further analysis is provided to identify by other statistical test if the pair of algorithms have significantly different performance. This is achieved with Tukey HSD test with confidence level of 95 % (reported at the bottom of Table 3), again the corrected p-value (0.0) give us a very strong presumption against null hypothesis.

5 Conclusions

We propose to integrate crossover operators in a dynamic island-based model for adaptive operator selection. This is implemented by using crossover with a similar formal signature to mutation, and keeping a temporary solution in the crossover island to serve as a parent. Importantly, our model is not a standard island model in that: (i) a single operator instead of complete evolutionary algorithm is kept in each island, and (ii) migration policies are dynamic rather than static. Our results on two benchmark problems (Onemax, and real-world instances of the course timetable problem), allow us to both prove the concept and test its practical relevance. Having a crossover island was found to significantly increase the performance, despite the added computational overhead.

Our results on the Onemax problem provide a visually appealing confirmation of an argument proposed by Ochoa et al. [13] on the advantages of recombination. Recombination performs a *dual-role* in genetic search according to the level of genetic diversity in the population. At early stages, when the population is diverse, recombination acts as a diverging operator (similar to a strong mutation), increasing the search power and speeding up the process. Towards the final stages of the search, when the population is genetically converged, recombination can instead focus the population around the fitness optimum (similar to a light mutation). Therefore, recombination has a dynamic role and is helpful across the complete search process.

Future work will explore the behaviour of more complex crossover operators and different migration policies over additional combinatorial problems.

Acknowledgments. J. A. Soria-Alcaraz would like to thank the *Consejo Nacional de Ciencia y tecnologia* (CONACyT, México). G. Ochoa would like to thank the University of Angers for hosting and funding a research visit in 2014 that started this collaboration.

References

1. Araujo, L., Guervós, J.J.M., Mora, A., Cotta, C.: Genotypic differences and migration policies in an island model. In: GECCO, pp. 1331–1338 (2009)
2. Burke, E.K., Gendreau, M., Hyde, M., Kendall, G., Ochoa, G., Ozcan, E., Qu, R.: Hyper-heuristics: a survey of the state of the art. J. Oper. Res. Soc. (JORS) **64**(12), 1695–1724 (2013)
3. Candan, C., Goëffon, A., Lardeux, F., Saubion, F.: A dynamic island model for adaptive operator selection. In: Genetic and Evolutionary Computation Conference (GECCO 2012), pp. 1253–1260 (2012)

4. DaCosta, L., Fialho, A., Schoenauer, M., Sebag, M.: Adaptive operator selection with dynamic multi-armed bandits. In: Proceedings of the 10th Annual Conference on Genetic and Evolutionary Computation, pp. 913–920. ACM (2008)
5. Fialho, Á., Da Costa, L., Schoenauer, M., Sebag, M.: Extreme value based adaptive operator selection. In: Rudolph, G., Jansen, T., Lucas, S., Poloni, C., Beume, N. (eds.) PPSN 2008. LNCS, vol. 5199, pp. 175–184. Springer, Heidelberg (2008)
6. Gustafson, S., Burke, E.K.: The speciating island model: an alternative parallel evolutionary algorithm. J. Parallel Distrib. Comput. **66**(8), 1025–1036 (2006)
7. Lardeux, F., Goëffon, A.: A dynamic island-based genetic algorithms framework. In: Deb, K., Bhattacharya, A., Chakraborti, N., Chakroborty, P., Das, S., Dutta, J., Gupta, S.K., Jain, A., Aggarwal, V., Branke, J., Louis, S.J., Tan, K.C. (eds.) SEAL 2010. LNCS, vol. 6457, pp. 156–165. Springer, Heidelberg (2010)
8. Lässig, J., Sudholt, D.: Design and analysis of migration in parallel evolutionary algorithms. Soft Comput. **17**(7), 1121–1144 (2013)
9. Luque, G., Alba, E.: Selection pressure and takeover time of distributed evolutionary algorithms. In: Pelikan, M., Branke, J. (eds.) Genetic and Evolutionary Computation Conference, GECCO 2010, pp. 1083–1088. ACM (2010)
10. Maturana, J., Saubion, F.: On the design of adaptive control strategies for evolutionary algorithms. In: Monmarché, N., Talbi, E.-G., Collet, P., Schoenauer, M., Lutton, E. (eds.) EA 2007. LNCS, vol. 4926, pp. 303–315. Springer, Heidelberg (2008)
11. McCollum, B., Schaerf, A., Paechter, B., McMullan, P., Lewis, R., Parkes, A.J., Gaspero, L.D., Qu, R., Burke, E.K.: Setting the research agenda in automated timetabling: the second international timetabling competition. INFORMS J. Comput. **22**(1), 120–130 (2010)
12. Ochoa, G., Hyde, M., Curtois, T., Vazquez-Rodriguez, J.A., Walker, J., Gendreau, M., Kendall, G., McCollum, B., Parkes, A.J., Petrovic, S., Burke, E.K.: HyFlex: a benchmark framework for cross-domain heuristic search. In: Hao, J.-K., Middendorf, M. (eds.) EvoCOP 2012. LNCS, vol. 7245, pp. 136–147. Springer, Heidelberg (2012)
13. Ochoa, G., Harvey, I., Buxton, H.: On recombination and optimal mutation rates. In: Proceedings of Genetic and Evolutionary Computation Conference (GECCO), pp. 488–495. Morgan Kaufmann (1999)
14. Rucinski, M., Izzo, D., Biscani, F.: On the impact of the migration topology on the island model. CoRR abs/1004.4541 (2010)
15. Skolicki, Z., Jong, K.D.: The influence of migration sizes and intervals on island models. In: GECCO, pp. 1295–1302 (2005)
16. Soria-Alcaraz, J., Martin, C., Héctor, P., Hugo, T.M., Laura, C.R.: Methodology of design: a novel generic approach applied to the course timetabling problem. In: Melin, P., Castillo, O. (eds.) Soft Computing Applications in Optimization, Control, and Recognition. Studies in Fuzziness and Soft Computing, vol. 294, pp. 287–319. Springer, Heidelberg (2013)
17. Soria-Alcaraz, J.A., Ochoa, G., Swan, J., Carpio, M., Puga, H., Burke, E.K.: Effective learning hyper-heuristics for the course timetabling problem. Eur. J. Oper. Res. **238**(1), 77–86 (2014)
18. Whitley, D., Rana, S., Heckendorn, R.: The island model genetic algorithm: on separability, population size and convergence. J. Comput. Inf. Technol. **7**, 33–47 (1998)

Parameter Setting for Multicore CMA-ES with Large Populations

Nacim Belkhir[1,2(✉)], Johann Dréo[1], Pierre Savéant[1], and Marc Schoenauer[2]

[1] Thales Research and Technology, Palaiseau, France
{nacim.belkhir,johann.dreo,pierre.saveant}@thalesgroup.com
[2] TAO, Inria Saclay Île-de-France, Orsay, France
marc.schoenauer@inria.fr

Abstract. The goal of this paper is to investigate on the overall performance of CMA-ES, when dealing with a large number of cores — considering the direct mapping between cores and individuals — and to empirically find the best parameter strategies for a parallel machine. By considering the problem of parameter setting, we empirically determine a new strategy for CMA-ES, and we investigate whether Self-CMA-ES (a self-adaptive variant of CMA-ES) could be a viable alternative to CMA-ES when using parallel computers with a coarse-grained distribution of the fitness evaluations. According to a large population size, the resulting new strategy for Self-CMA-ES and CMA-ES, is experimentally validated on BBOB benchmark where it is shown to outperform a CMA-ES with default parameter strategy.

Keywords: Empirical study · Numerical optimization · Metaheuristics · Algorithms comparison

1 Introduction

Covariance Matrix Adaptation Evolution Strategy (CMA-ES) [6] is one of the most efficient algorithms for real valued single-objective optimization problems. Thanks to its invariance properties [10], some default parameter values could be tuned using a rather small set of test functions [6], and nevertheless provide robust performances on a large variety of problems, from analytical benchmark functions [8] to many real-world applications (see, among many others, [7]).

With the end of Moore's years, increasing the speed of software nowadays requires an efficient parallelisation. Evolutionary Algorithms like CMA-ES can trivially be parallelized without modifying the underlying dynamics of the algorithm by distributing the computation of the fitnesses of the whole population on different slave nodes, the master node maintaining the population as a whole, and ensuring the reproduction phase. For optimal efficiency, the population size should be some multiple of the number of available computing units.

It turns out that the default value for the population size λ for CMA-ES is rather small, empirically set to $4 + 3 \ln(n)$ [6], where n is the problem dimension.

© Springer International Publishing Switzerland 2016
S. Bonnevay et al. (Eds.): EA 2015, LNCS 9554, pp. 109–122, 2016.
DOI: 10.1007/978-3-319-31471-6_9

And increasing λ without any further parameter tuning has been experimentally demonstrated to perform poorly for CMA-ES and other types of Evolution Strategies: [3] proposes a new update strategy for the global step-size; [21,22] suggests to modify the ratio between number of parents and number of offspring. This paper investigates another approach to improve the performance of CMA-ES in a distributed setting: assuming some given number of cores, the use of computing resources is optimized by fixing the population size λ to this number of cores[1]. The goal is then to optimize the other parameters of CMA-ES to improve its performances.

Today, parameter tuning is acknowledged as a mandatory step toward efficient optimization algorithms at large [11], be they exact combinatorial optimization algorithms [12], or (possibly stochastic) heuristics and metaheuristics, among which Evolutionary Algorithms [5] (more in Sect. 2.1). Off-line tuning considers parameter tuning as a (meta-)optimization problem, and generic optimization algorithms can hence be applied [4,12,14,18]. These methods have been in particular used to further improve CMA-ES performances [13,15,19], therefore suggesting that the same approach could be used to tackle the problem of a large λ – though leaving open the issue of the generality of such tuning [20].

On the other hand, optimization is a dynamic process, and the best parameter values at a given time of the search might no longer be efficient later. On-line parameter tuning therefore seems a very promising approach. However, there are very few (if any) examples of success of on-line tuning except in the history of Evolution Strategies, where CMA-ES, as its name suggests, is the most sophisticated of a long line of algorithms that do efficiently implement on-line adaptation of their main parameters. Yet, the adaptation mechanism of CMA-ES itself has some parameters, and a first approach to their on-line tuning has been recently proposed, leading to the so-called Self-CMA-ES [17], validated on a few test functions, and in the framework of a large population size.

The goal of this work is to investigate CMA-ES parameter tuning in a distributed context (fixed large λ), and in particular to compare experimentally the off-line and on-line approaches for different values of λ on the BBOB benchmark suite. The paper is organized as follows. Section 2 rapidly introduces the problem of parameter setting, and details the hyper-parameters of CMA-ES and how Self-CMA-ES adapts them. Section 3 introduces the experimental protocol that is used in Sect. 4.4 to validate some choices of Self-CMA-ES and compare the different approaches. Finally, the results are discussed and further research directions are proposed in Sect. 5.

2 State of the Art

2.1 Parameter Setting

It is today widely acknowledged that the performances of optimization algorithms are highly correlated with the values given to their parameters [11]. Following the classification discussed in [5], one should distinguish between off-line

---

[1] This also covers the case where λ is set to some multiple of the number of cores.

and on-line parameter setting methods. In the off-line case (aka parameter tuning), the important secondary issue is that of the generality of the setting, and in the on-line case (aka parameter control), the distinction between dynamic, adaptive or self-adaptive approaches.

Off-line approaches view the problem of parameter tuning as an optimization problem in the space of parameters: the fitness of a parameter setting is the performance of the algorithm at hand, and any optimization method on the parameter space can be used given a practical way to compute the performance of the algorithm. Assuming that the user knows the quantity she/he is interested in (e.g., minimizing the runtime to reach a given solution quality, or optimizing the solution quality given a fixed computational budget), here comes into play the generality of the sought setting [20]. If the target of the experiments is a single (or a small number of) problem instance(s), the performance of the algorithm is computed by running it on each target instance (eventually aggregating over the different instances in the target set). But very often, the goal of parameter tuning is to find a robust setting that will give very good performances for some class of problem instances that cannot be enumerated. The performance of the algorithm is then approximated by running it on some carefully chosen test set of instances of the target class, hoping the result will be general enough to cover the whole class. Using large test sets improves the robustness of the setting, but increases the computational cost of the parameter setting process, as one single evaluation of the performance of a given parameter setting involves running the algorithm at hand once for all instances of the test set.

Several generic optimization methods have been adapted to handle parameter tuning and cope with the above-mentioned generalization issue, based on racing [16], on metaheuristics [18], on statistical modeling of the algorithm performance with Gaussian Processes [2], or on local search [14]. The most recent one, that has been used in this work, is SMAC (Sequential Model-based Algorithm Configuration)[2] [12], that uses random forest regression to model the algorithm performance as well as the uncertainty of its prediction. SMAC uses the Expected Improvement measure to choose, given a model, which parameter set to try next.

On-line parameter control, on the other hand, is concerned with tuning the parameter values during the run of the algorithm, thus avoiding any generalization issue and, more importantly, requiring little, if any, computational overhead. Three approaches should be distinguished [5], depending on how the parameters are modified during the run: in the deterministic approach, they are modified using a fixed schedule (that has to be designed off-line!); in the adaptive approach, the parameters are modified according to some feedback from the current state of the search; and in the self-adaptive approach, the parameters are subject to evolution: each individual (potential solution of the original optimization problem) carries its own parameters, and though selection applies only to the fitness, it is hoped that successive selections will only select individuals that carry good parameters.

---

[2] SMAC is freely available at http://www.cs.ubc.ca/labs/beta/Projects/SMAC/.

Unfortunately, whilst adaptive or self-adaptive on-line control is potentially more efficient than off-line tuning, offering a way to adapt the parameters to the instance at hand, and to the current state of the search, there are very few examples of successful on-line control, and most of them are highly problem-dependent. As a matter of fact, the only success story of on-line parameter tuning is that of Evolution Strategies. A detailed presentation of the history of Evolution Strategies in this perspective is given in Sect. 3 of [5] and will not be repeated here due to space restrictions. We will directly switch to introducing CMA-ES, that can be viewed as the last link of the long chain of Evolution Strategies variants, that went from adaptive to self-adaptive and back to adaptive tuning of its Gaussian mutation.

2.2 CMA-ES

Let f be the real-valued objective function, defined on $\mathbb{R}^n$. CMA-ES [6] evolves a Gaussian distribution $\mathcal{N}\left(\boldsymbol{m}^t, (\sigma^t)^2 \boldsymbol{C}^t\right)$ on $\mathbb{R}^n$ with mean $\boldsymbol{m}^t$ (the current estimate of the optimal solution) and covariance matrix $(\sigma^t)^2 \boldsymbol{C}^t$, where the step-size σ^t is isolated from the covariance direction $\boldsymbol{C}$ so they can be adapted separately. The original $(\mu/\mu_w, \lambda)$-CMA-ES (Algorithm 1) works as follows. At iteration t, the current distribution $\mathcal{N}\left(\boldsymbol{m}^t, (\sigma^t)^2 \boldsymbol{C}^t\right)$ is sampled, generating λ candidate solutions (line 5), whose fitness is computed (line 6). The new mean $\boldsymbol{m}^{t+1}$ is computed line 7 as the weighted sum of the best μ individuals according to f. The adaptation of the step-size σ^t is controlled by the evolution path $\boldsymbol{p}_\sigma^{t+1}$, that stores, with relaxation factor c_σ, the successive mutation steps $\frac{\boldsymbol{m}^{t+1}-\boldsymbol{m}^t}{\sigma^t}$ (line 8). The step-size is increased (resp. decreased) in the case of the length of the evolution path $\boldsymbol{p}_\sigma^{t+1}$ is longer (resp. smaller) than the expected length it would have under random selection (line 9). The covariance matrix is updated using both a rank-one update term, computing the evolution path $\boldsymbol{p}_c^{t+1}$ of successful moves of the mean $\frac{\boldsymbol{m}^{t+1}-\boldsymbol{m}^t}{\sigma^t}$ of the distribution in the original coordinate system (line 11) and the rank-μ update, a weighted sum of the covariances of successful steps of the best μ individuals (using the weights of the update of the mean – line 12). Two weights are used for this last update (line 13), c_1 for the rank-one term, and c_μ for the rank-μ term, hence c_1 and c_μ must be positive with $c_1 + c_\mu \leq 1$.

The default values of the parameters of the algorithm [6] are set in line 1, but are hidden to the user in the standard CMA-ES distributions, except for the population size λ and the number of selected parents μ. Though the already-mentioned invariance properties of CMA-ES [10] ensure some robutsness of the default setting, several improvements could be reached using off-line tuning of some of these parameters, namely λ (or more precisely the coefficient of λ as a function of n) and the ratio $\frac{\mu}{\lambda}$, as well as the parameters c_σ and d_σ for the adaptation of σ [13,19]. Note that some additional parameters related to the stopping criterion are not presented in Algorithm 1, and have a large impact on the restart versions of CMA-ES [1]. These were also tuned using IRACE in [15]. However, to the best of our knowledge, the parameter setting for the adaptation of the covariance matrix c_c (line 11, c_1 and c_μ (line 13) has only been addressed on-line in [17], and will now be detailed.

Algorithm 1. The $(\mu/\mu_w, \lambda)$-CMA-ES (from [6])

1: **given** $n \in \mathbb{N}_+$, $\lambda = 4 + \lfloor 3 \ln n \rfloor$, $\mu = \lfloor \lambda/2 \rfloor$, $w_i = \frac{\ln(\mu + \frac{1}{2}) - \ln i}{\sum_{j=1}^{\mu}(\ln(\mu + \frac{1}{2}) - \ln j)}$ for $i = 1 \ldots \mu$,

$\mu_w = \frac{1}{\sum_{i=1}^{\mu} w_i^2}$, $c_\sigma = \frac{\mu_w + 2}{n + \mu_w + 3}$, $d_\sigma = 1 + c_\sigma + 2\max(0, \sqrt{\frac{\mu_w - 1}{n+1}} - 1)$, $c_c = \frac{4}{n+4}$,

$c_1 = \frac{2}{(n+1.3)^2 + \mu_w}$, $c_\mu = \frac{2(\mu_w - 2 + 1/\mu_w)}{(n+2)^2 + \mu_w}$

2: **initialize** $m^{t=0} \in \mathbb{R}^n$, $\sigma^{t=0} > 0$, $p_\sigma^{t=0} = 0$, $p_c^{t=0} = 0$, $C^{t=0} = I$, $t \leftarrow 0$

3: **repeat**

4:     **for** $k = 1, \ldots, \lambda$ **do**

5:       $x_k = m^t + \sigma^t \mathcal{N}(0, C^t)$

6:       $f_k = f(x_k)$

7:     $m^{t+1} = \sum_{i=1}^{\mu} w_i x_{i:\lambda}$

8:     $p_\sigma^{t+1} = (1 - c_\sigma) p_\sigma^t + \sqrt{c_\sigma(2 - c_\sigma)} \sqrt{\mu_w} (C^t)^{-\frac{1}{2}} \frac{m^{t+1} - m^t}{\sigma^t}$

9:     $\sigma^{t+1} = \sigma^t \exp(\frac{c_\sigma}{d_\sigma}(\frac{\|p_\sigma^{t+1}\|}{\mathbb{E}\|\mathcal{N}(0,I)\|} - 1))$

10:    $h_\sigma = \mathbb{1}_{\|p_\sigma^{t+1}\| < \sqrt{1 - (1 - c_\sigma)^{2(t+1)}}(1.4 + \frac{2}{n+1})\mathbb{E}\|\mathcal{N}(0,I)\|}$

11:    $p_c^{t+1} = (1 - c_c) p_c^t + h_\sigma \sqrt{c_c(2 - c_c)} \sqrt{\mu_w} \frac{m^{t+1} - m^t}{\sigma^t}$

12:    $C_\mu = \sum_{i=1}^{\mu} w_i \frac{x_{i:\lambda} - m^t}{\sigma^t} \times \frac{(x_{i:\lambda} - m^t)^T}{\sigma^t}$

13:    $C^{t+1} = (1 - c_1 - c_\mu) C^t + c_1 \underbrace{p_c^{t+1} p_c^{t+1 T}}_{\text{rank-one update}} + c_\mu \underbrace{C_\mu}_{\text{rank-}\mu\text{ update}}$

14:    $t = t + 1$

15: **until** *stopping criterion is met*

2.3 Self-CMA-ES

In Self-CMA-ES [17], the on-line tuning of c_c, c_1, c_μ relies on the hypothesis that the best parameter configuration at time t is the one that would have maximized at time $t - 1$ the likelihood of generating the best individuals selected at time t. At every iteration t, an auxiliary optimization algorithm (another CMA-ES, denoted CMA-ES$_{aux}$) is hence used to compute this optimal configuration. After computing the λ offspring at time t (lines 4–5 of Algorithm 1), the state of the algorithm at time $t - 1$ is restored, and the optimization of parameters c_c, c_1, c_μ proceeds as follows: for each triplet value (c_c, c_1, c_μ), the virtual distribution parameters σ and C are computed (lines 8–13) from state $t - 1$, and the performance of (c_c, c_1, c_μ) is the likelihood of generating the best μ of the actual λ offspring at time t from this virtual distribution. The triplet (c_c, c_1, c_μ) that maximizes this likelihood is returned and is then used, at time t, to complete the actual update of the actual mutation parameters of CMA-ES (lines 8–13).

A first issue is that computing the log-likelihood of generating μ given points of $\mathbb{R}^n$ from a given Gaussian is costly and numerically unstable. It was hence replaced by a proxy, that works as follows. λ points are sampled from the virtual Gaussian, their virtual mean is computed (as in line 7), and the Mahalanobis distance between the actual μ best offspring at time t and this mean is computed. The sum of ranks of these distances used as a proxy for the likelihood.

The detailed formal description of this proxy for the likelihood is given in [17], together with the global Self-CMA-ES algorithm.

A second issue is the possible overfitting of the parameters (c_c, c_1, c_μ) due to a single and limited sampling of the actual offspring at time t. And a third issue is the computational cost of running a full CMA-ES$_{aux}$ inside every iteration of the master CMA-ES: even though no additional fitness computation of the main CMA-ES is required, and even though the dimension of the auxiliary optimization problem is only 3, sampling the virtual Gaussian distribution to evaluate the proxy likelihood of many triples (and here the dimension is n) has a non-negligible cost. However, both issues can be resolved simultaneously. First, the CMA-ES$_{aux}$ is not restarted from scratch at every iteration t of the main CMA-ES, but restarts from the state of the CMA-ES$_{aux}$ at the end of iteration $t - 1$; Second, only a small number of iterations of CMA-ES$_{aux}$ is actually run, avoiding possible overfitting. Section 4.1 will describe some experimental validation of this procedure.

3   Experimental Setting

The remainder of the paper is devoted to presenting experimental comparison with the goal of validating some choices for Self-CMA-ES, and assessing when and how Self-CMA-ES is a better choice than CMA-ES with its default values.

BBOB testbench. All experiments use test functions from the Black Box Optimization Benchmark (BBOB)[3] [9]. BBOB testbench contains 24 functions, with known difficulty (e.g. non-separability, high conditioning, different levels of multimodality, with or without global structure, etc.) and for different dimensions (2, 3, 5, 10, 20, 40). BBOB also proposes an API for most programming languages. To avoid any bias, for each function, 15 trials are run, where for each trial, the optimum is moved and for the non-separable functions, the coordinate system is rotated. Foreach trial, a maximum number of function evaluations of $10^5 * n$ is given before the algorithm is killed. Only the noiseless versions of the functions were used here.

Performance Measure. BBOB uses as performance measure the Expected Run Time, that counts the number of function evaluations used to reach a given target objective value taking into account the runs that failed to reach that target value. This computational effort is normalized by dividing it by the dimension, when the results on different dimensions need to be aggregated. In this work, we only consider one target value 10^{-8}, and the number of function evaluations #FEs as a measure of comparison. However, because we are interested in the distributed performance, in a context where only the time-to-solution matters, we propose a new performance measure, the **Virtual Wall Clock Time** ($VWCT$), focusing on the core usage, and formally defined as:

$$VWCT = \frac{\#\text{FEs}}{\lambda} = \frac{\#\text{FEs}}{\#\text{cores}} \tag{1}$$

---

[3] http://coco.gforge.inria.fr.

The communication time is here neglected: in real situations on HPC clusters, it will be several orders of magnitude smaller than the computation time of the objective function (even if this is not true for BBOB functions).

Implementation. For all experiments, we used the Octave/MATLAB source code provided by authors of [17][4], that was modified in order to expose the parameters for automated parameter tuning, and/or to apply new parameter strategies to some parameters.

4   Experimental Results

Four series of experiments are conducted. A first goal is to validate some choices made in [17] for Self-CMA-ES; A second goal is to compare Self-CMA-ES with some off-line tuning of (c_c, c_1, c_μ); A third goal is to identify the best strategy for the choice of μ; and the final goal is to assess on the whole BBOB benchmark suite, the performances of Self-CMA-ES with respect to CMA-ES (using the best setting that could be deduced from the previous experiments).

4.1   Validation of Self-CMA-ES

A first sanity check of Self-CMA-ES is performed by tuning the initial values of c_c, c_1, c_μ with SMAC. The good news is that the performance of Self-CMA-ES is not sensitive to these initial values, as the adaptive mechanism takes over, whatever its initialization.

A second experiment checks the strategy for CMA-ES$_{aux}$ (see Sect. 2.3), running it for different number of iterations, or to full completion. The clear conclusion is that indeed, as argued in [17], the best results are obtained when running a single iteration of CMA-ES$_{aux}$ at each iteration of the main CMA-ES. Because of the space constraints, none of these validation experiments is detailed here.

4.2   On-Line vs Off-Line Tuning of c_c, c_1, c_μ

In order to check the efficiency of the on-line tuning of c_c, c_1, c_μ done by Self-CMA-ES, it should be compared to the off-line tuning of the same parameters (e.g., using SMAC, see Sect. 2.1) on the plain CMA-ES. However, because it was demonstrated in [3,21] that the performance of CMA-ES (or other Evolution Strategies) with a large λ was highly dependent on μ and the adaptation of σ, and also because SMAC experiments are very costly, it was decided to run one single SMAC campaign, tuning μ and σ_0, the initial value for σ, for both algorithms (using the adaptation scheme advocated in [3,21] is left for further work), and c_c, c_1, c_μ for CMA-ES. Table 1 describes the experimental conditions. Note additionally that c_1 and c_μ must satisfy an additional constraint, that was handled by returning a very high fitness without running the algorithm when violated.

---

[4] https://sites.google.com/site/selfcmappsn/.

Table 1. Experimental setting for SMAC on CMA-ES and Self-CMA-ES.

Test Functions	F1-Sphere, F8-Rosenbrock, F13-Sharp Ridge, F16-Rastrigin
Dimensions	10, 20
λ	λ_{def}, 50, 100, 200, 500, 1000, 1500, 2000
SMAC target for Self-CMA-ES	$\mu \in [1, \lambda]$, $\sigma_0 \in [0, 2]$
SMAC target for CMA-ES	$\mu \in [1, \lambda]$, $\sigma_0 \in [0, 2]$, $(c_c, c_1, c_\mu) \in [0, 1]^3$

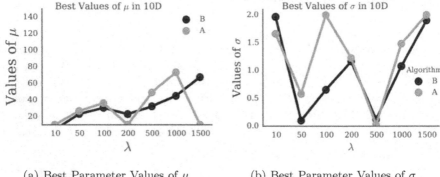

(a) Best Parameter Values of μ (b) Best Parameter Values of σ

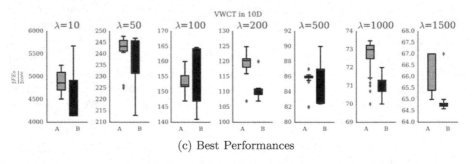

(c) Best Performances

Fig. 1. Results for SMAC (see Table 1): Best values for μ (a) and σ_0 (b), and best performances (c) of CMA-ES (A) and Self-CMA-ES (B) on 10D-Rosenbrock.

Typical results are given in Fig. 1. The best values for μ (Fig. 1a) are in agreement with [22], i.e., are lower than the default $\frac{\lambda}{2}$. Some regularity with respect to λ could however be identified, and will be investigated in Sect. 4.3.

Figure 1b is typical of the behavior of the best values of σ_0. Apart the fact that they usually are lower than the value used in [17] (2.0), it was not possible to fit any relation with the dimention of the problem. However, the influence of this parameter seemed limited accross the experiments. Hence, all further experiments will use $\sigma = 1.3$, a rough average of all best values returned by SMAC.

No trend could be observed either for c_c, c_1, c_μ, except a rather large variance of the best values returned by SMAC. Thus the default parameter setting [6] will be used in the remaining of the experiments for CMA-ES.

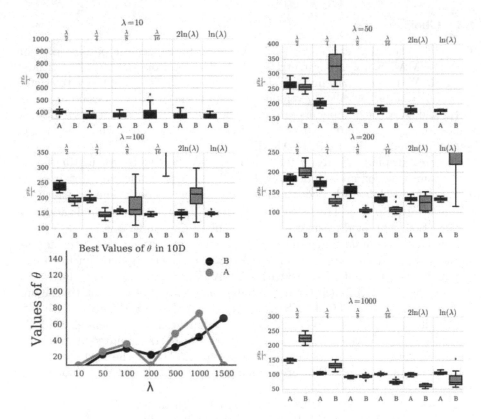

Fig. 2. Performances of CMA-ES (A, black) and Self-CMA-ES (B, grey) on 10D-Attractive Sector ($\lambda_{def} = 10$), for all pairs (λ, μ) of Table 2. Empty columns mean poor results (scaled for readability).

Table 2. Setting for the "μ" experiments. σ_0 is set to 1.3.

	Values
λ	(λ_{def}, 50, 100, 150, 200, 500, 1000)
μ	$\frac{\lambda}{2}$, $\frac{\lambda}{8}$, $\frac{\lambda}{16}$, $2 * ln(\lambda)$, $ln(\lambda)$
Functions	F1-Sphere, F6-Attractive Sector, F8-Rosenbrock,
	F11-Discuss, F12-Bent Cigar
Dimensions	2, 10, 20

Finally, Fig. 1c plots the best overall values of both algorithms using the best parameterization returned by SMAC for each of them. The good news is that for all λ, Self-CMA-ES can be tuned to perform at least as good as the best tuning of CMA-ES, though the large variances suggest that more experiments should be run to better assess this conclusion.

4.3 Choice of μ

The goal of the next series of experiments is to find a generic parametrization for Self-CMA-ES, i.e. a parametrization that is good on all instances without using SMAC for each new instance. The possible values for μ are hence restricted to the discrete list of values given on Table 2, depending on λ. As said, σ_0 is set to 1.3 and all other parameters are set to the default value. As for CMA-ES, the values for c_c, c_1, c_μ are set to their default values as well – while they are of course adapted on-line by Self-CMA-ES.

Figure 2 displays the result for function F6-Attractive Sector in 10D for all (λ, μ) pairs. As for all functions of Table 2, the best values are obtained for $\mu \in [\frac{\lambda}{4}, \ln(\lambda)]$, while both algorithms achieve their worst performances whith the default strategy $\mu = \frac{\lambda}{2}$. The value $\mu = \frac{\lambda}{8}$ is hence retained for the final validation next Section, as providing quasi-optimal results for all functions.

Yet another validation of the on-line strategy for setting c_c, c_1, c_μ is presented on Fig. 3, that compares, for $\lambda = 200$, and on the F1-Sphere function on 10 and 20 dimensions, Self-CMA-ES with a CMA-ES for which c_c, c_1, c_μ have been tuned using SMAC for each value of μ independently, denoted A* on the Figure. The results of the tuned CMA-ES are better than those of Self-CMA-ES, though not significantly for the chosen value $\mu = \frac{\lambda}{8}$. Furthermore, remember that the tuning with SMAC requires to run the algorithm several hundreds times. Furthermore, applying the parameters returned by SMAC for the 20D case to the 10D case displays results that are similar to those of Self-CMA-ES (not shown here).

Another interesting conclusion that can be drawn from Fig. 2 is that the $VWCT$ for $\lambda = 500$ and $\lambda = 1000$ have very similar values: adding more cores does not help, and other strategies are needed to take full benefit of CMA-ES on large computing clusters.

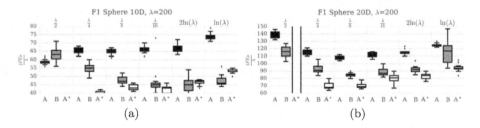

Fig. 3. Comparison of CMA-ES (A, black), Self-CMA-ES (B, grey), and the tuned CMA-ES (A*, white) on the Sphere for $\lambda = 200$ and as in μ of Table 2.

4.4 Overall BBOB Comparisons

The final experiment is to perform complete BBOB comparisons between the retained generic parametrization for both Self-CMA-ES and CMA-ES, i.e.,

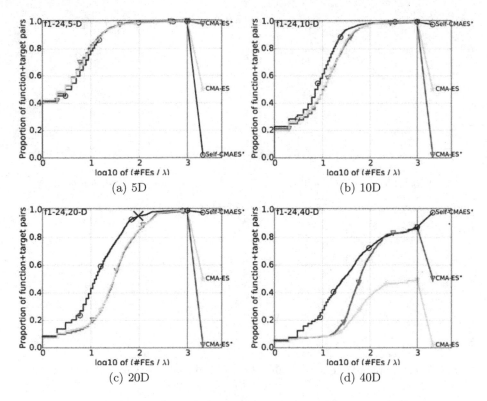

Fig. 4. Bootstrapped empirical cumulative distribution of $VWCT$ for all functions in 5D, 10D, 20D, 40D. Self-CMA-ES$^{\bullet}$ and CMA-ES$^{\bullet}$ use $\mu = \frac{\lambda}{8}$ while CMA-ES uses the default $\mu = \frac{\lambda}{2}$.

$\mu = \frac{\lambda}{8}$ and $\sigma_0 = 1.3$ (the values of c_c, c_1, c_μ are set to their default values for CMA-ES). The curve for the default strategy for CMA-ES ($\mu = \frac{\lambda}{8}$) was also added to the comparison. The case $\lambda = 200$ was chosen as representative.

Figure 4 displays the aggregated results for all functions, for dimensions 5, 10, 20 and 40. Except for dimension 5, Self-CMA-ES performs better than both CMA-ES, and the advantage increases with the dimension. Looking now at more detailed results, on Fig. 5, it can be seen that the worst results for Self-CMA-ES are obtained for the separable functions in dimension 40 (Fig. 5a), where it sometimes fails to reach the target value. Further investigations are needed to understand why this happens. Also note that the results for low or moderate conditioning functions (not shown here) show slightly worse results (though not as bad as for the separable functions) for Self-CMA-ES.

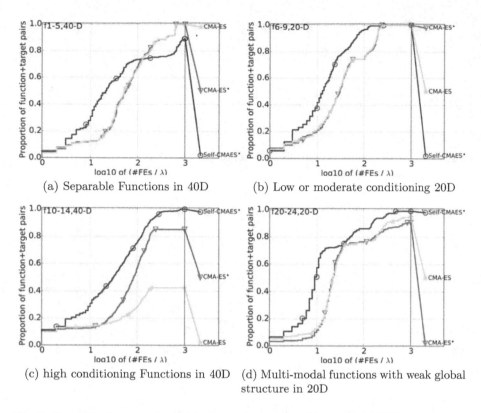

(a) Separable Functions in 40D   (b) Low or moderate conditioning 20D

(c) high conditioning Functions in 40D   (d) Multi-modal functions with weak global structure in 20D

Fig. 5. Some Bootstrapped empirical cumulative distribution of *VWCT*. Legend as in Fig. 4.

5   Discussion and Conclusion

This paper has experimentally studied parametrization strategies of CMA-ES that is run in a distributed environment when the primary goal is to minimize the wall-clock time-to-solution by using all available computing units (e.g., cores). This situation was simulated by considering large values of the population size λ as constraints, and tuning the other parameters accordingly. In particular, the Self-CMA-ES approach [17] has been demonstrated to be, in most cases, a viable alternative to the default CMA-ES for that goal.

The experiments presented in this paper have first validated most of the choices made in the original Self-CMA-ES approach [17] for the online control of the usually hidden parameters c_c, c_1, c_μ that govern the update of the covariance matrix in CMA-ES. For values of λ up to 2000, we have observed that the best strategy for the choice of the number of parents μ is $\mu = \frac{\lambda}{8}$. This strategy outperforms the default strategy $\mu = \frac{\lambda}{2}$, for both CMA-ES and Self-CMA-ES. Also, this new strategy slightly outperforms the strategy $\mu = \frac{\lambda}{4}$ defined in [22], although [22] considers larger values of λ.

Regarding the initial value σ_0 for the step-size σ, the best value for both CMA-ES and Self-CMA-ES was found to be smaller than that used in [17] ($\sigma = 2$), while nevertheless higher than the default value used [6] ($\sigma = 0.3$). The latter is explained by the increase of λ, resulting in a larger coverage of the search space by the initial sampling. Additionally, the new value of σ asserts the assumption of adapting the step-size when dealing with a larger λ, as proposed in [21].

The resulting new strategy for Self-CMA-ES and CMA-ES uncovers good performances, significantly outperforming the default strategy. Moreover, even when CMA-ES is tuned off-line anew for each problem instance, Self-CMA-ES remains a good alternative to CMA-ES, performing only slightly worse while avoiding the huge computational cost of the tuning process.

More work is needed, however, in order to take full benefit of a very large number of computing units, as the wall-clock time performance seems to stagnate above 500 cores. Possible directions are to hybridize the method proposed here with those proposed in [3,21] and also modify the adaptation mechanism of the step-size σ. Another further direction is concerned with detecting situations where Self-CMA-ES adaptation mechanism performs poorly, and to switch back to the default values for c_c, c_1, c_μ in such cases, thus guaranteeing performances at least as good as those of CMA-ES. Another approach would be to consider a portfolio of strategies in order to maximize the expected performance of CMA-ES, that should include CMSA-ES [3], that outperforms CMA-ES in large dimensions and population sizes.

References

1. Auger, A., Hansen, N.: A restart CMA evolution strategy with increasing population size. In: CEC 2005, vol. 2, pp. 1769–1776. IEEE (2005)
2. Bartz-Beielstein, T., Lasarczyk, C.W., Preuß, M.: Sequential parameter optimization. In: CEC 2005, vol. 1, pp. 773–780. IEEE (2005)
3. Beyer, H.-G., Sendhoff, B.: Covariance matrix adaptation revisited – The CMSA evolution strategy –. In: Rudolph, G., Jansen, T., Lucas, S., Poloni, C., Beume, N. (eds.) PPSN 2008. LNCS, vol. 5199, pp. 123–132. Springer, Heidelberg (2008)
4. Birattari, M., Stützle, T., Paquete, L., Varrentrapp, K., et al.: A racing algorithm for configuring metaheuristics. In: Langdon, W.B., et al. (ed.) Proceedings of ACM GECCO 2002, pp. 11–18 (2002)
5. Eiben, A., Michalewicz, Z., Schoenauer, M., Smith, J.E.: Parameter control in evolutionary algorithms. In: Lobo, F., Lima, C.F., Michalewicz, Z. (eds.) Parameter Setting in Evolutionary Algorithms, pp. 19–46. Springer, Heidelberg (2007)
6. Hansen, N., Müller, S., Koumoutsakos, P.: Reducing the time complexity of the derandomized evolution strategy with covariance matrix adaptation (CMA-ES). Evol. Comput. **11**(1), 1–18 (2003)
7. Hansen, N., Niederberger, S., Guzzella, L., Koumoutsakos, P.: A method for handling uncertainty in evolutionary optimization with an application to feedback control of combustion. IEEE Trans. Evol. Comput. **13**(1), 180–197 (2009)
8. Hansen, N.: Benchmarking a BI-population CMA-ES on the BBOB-2009 function testbed. In: Rothlauf, F. (ed.) GECCO Companion, pp. 2389–2396. ACM (2009)

9. Hansen, N., Auger, A., Finck, S., Ros, R.: Real-parameter black-box optimization benchmarking 2010: experimental setup. Technical report RR-7215, INRIA (2010)
10. Hansen, N., Ros, R., Mauny, N., Schoenauer, M., Auger, A.: Impacts of invariance in search: when CMA-ES and PSO face Ill-conditioned and non-separable problems. Appl. Soft Comput. **11**, 5755–5769 (2011)
11. Hoos, H.H.: Programming by optimization. Commun. ACM **55**(2), 70–80 (2012)
12. Hutter, F., Hoos, H.H., Leyton-Brown, K.: Sequential model-based optimization for general algorithm configuration. In: Coello, C.A.C. (ed.) LION 2011. LNCS, vol. 6683, pp. 507–523. Springer, Heidelberg (2011)
13. Hutter, F., Hoos, H.H., Leyton-Brown, K., Murphy, K.P.: An experimental investigation of model-based parameter optimisation: SPO and beyond. In: Rothlauf, F. (ed.) GECCO 2009, pp. 271–278. ACM (2009)
14. Hutter, F., Hoos, H.H., Leyton-Brown, K., Stützle, T.: ParamILS: an automatic algorithm configuration framework. JAIR **36**(1), 267–306 (2009)
15. Liao, T., Stützle, T.: Testing the impact of parameter tuning on a variant of IPOP-CMA-ES with a bounded maximum population size on the noiseless BBOB testbed. In: Proceedings of ACM GECCO, pp. 1169–1176. ACM (2013)
16. López-Ibáñez, M., Dubois-Lacoste, J., Stützle, T., Birattari, M.: The R-package Irace, iterated race for automatic algorithm configuration. Technical report TR/IRIDIA/2011-004, IRIDIA, Université Libre de Bruxelles, Belgium (2011)
17. Loshchilov, I., Schoenauer, M., Sebag, M., Hansen, N.: Maximum likelihood-based online adaptation of hyper-parameters in CMA-ES. In: Bartz-Beielstein, T., Branke, J., Filipič, B., Smith, J. (eds.) PPSN 2014. LNCS, vol. 8672, pp. 70–79. Springer, Heidelberg (2014)
18. Nannen, V., Eiben, A.E.: Relevance estimation and value calibration of evolutionary algorithm parameters. In: IJCAI 2007, vol. 7, pp. 6–12 (2007)
19. Smit, S., Eiben, A.: Beating the "World champion" evolutionary algorithm via REVAC tuning. In: Proceedings of IEEE Congress on Evolutionary Computation, pp. 1–8, July 2010
20. Smit, S.K., Eiben, A.E.: Parameter tuning of evolutionary algorithms: generalist vs. specialist. In: Di Chio, C., et al. (eds.) EvoApplicatons 2010, Part I. LNCS, vol. 6024, pp. 542–551. Springer, Heidelberg (2010)
21. Teytaud, F.: A new selection ratio for large population sizes. In: Di Chio, C., et al. (eds.) EvoApplicatons 2010, Part I. LNCS, vol. 6024, pp. 452–460. Springer, Heidelberg (2010)
22. Teytaud, F., Teytaud, O.: On the parallel speed-up of estimation of multivariate normal algorithm and evolution strategies. In: Giacobini, M., et al. (eds.) EvoWorkshops 2009. LNCS, vol. 5484, pp. 655–664. Springer, Heidelberg (2009)

Towards Human-Competitive Game Playing for Complex Board Games with Genetic Programming

Denis Robilliard$^{(\boxtimes)}$ and Cyril Fonlupt

LISIC, ULCO, University of Lille–Nord de France, Calais, France
denis.robilliard@lisic.univ-littoral.fr

Abstract. Recent works have shown that Genetic Programming (GP) can be quite successful at evolving human-competitive strategies for games ranging from classic board games, such as chess, to action video games. However to our knowledge GP was never applied to modern complex board games, so-called eurogames, such as Settlers of Catan, i.e. board games that typically involve four characteristics: they are non zero-sum games, multiplayer, with hidden information and random elements. In this work we study how GP can evolve artificial players from low level attributes of a eurogame named "7 Wonders", that features all the characteristics of this category. We show that GP can evolve competitive artificial intelligence (AI) players against human-designed AI or against Monte Carlo Tree Search, a standard in automatic game playing.

1 Introduction

Games are a classic AI research subject, with well-known successful results on games like chess, checkers, or backgammon. However, complex board-games, also nicknamed eurogames, still constitute a challenge, which has been initiated by such works as [1] or [2] on the game "Settlers of Catan", or [3] on the game "Dominion". Most often these games combine several characteristics among being no zero-sum games, multiplayer, with incomplete information and random elements, together with little formalized expert knowledge on the subject. Monte-Carlo Tree Search (MCTS), which gained much notoriety from the game of Go [4], seems a method of interest in this context because it does not require background knowledge about the game. Genetic Programming (GP) could also qualify as a possible alternative, for the very same reason.

In their pioneering work [5], Hauptman and Sipper were successful at developing winning strategies for chess endgames with GP, and Sipper also achieved impressive results on Robocode and Backgammon [6]. For instance the top evolved strategy for Backgammon was able to get a win percentage of 62.4 % in a tournament against Pubeval, one of the strongest linear neural network player. However to the best of our knowledge, GP has never been used to automatically evolve a competitive player for a complex eurogame.

In order to simplify the obtaining of an AI for eurogames, many published works (whatever the AI engine) use some restricted game configuration or only a

© Springer International Publishing Switzerland 2016
S. Bonnevay et al. (Eds.): EA 2015, LNCS 9554, pp. 123–135, 2016.
DOI: 10.1007/978-3-319-31471-6_10

limited subset of the game rules. E.g. in [2] no trade interactions between players are allowed, and in [3] only a subset of the possible cards are used.

In this paper we focus on the creation of a GP evolved AI player for the "7 Wonders" (7W) eurogame, presented in the next section. In order to test the potential of GP, one of our requirements is to tackle the full game rules, including the trading mechanism.

In the next section we introduce the 7W game and its rules. Then we present the program architecture that was used for evolving GP players. After dealing with specific issues that arose during implementation or testing, we present our experiments and their results.

2   Description of the "7 Wonders" Game

Board games are increasingly numerous, with more than 500 games presented each year at the international Essen game fair. Among these, the game "7 Wonders" (7W), issued in 2011, obtains a fair amount of success, with about 100,000 copies sold per year and several expansion sets. It is basically a card game, whose theme is similar to the many existing "civilization" computer games, where players develop a virtual country using production of resources, trade, military and cultural improvements.

The 7W game belongs in the family of partially observable, multiplayer, stochastic, and also competitive games (although in any N-player game with $N > 2$, several players may share cooperative sub-goals, such as hindering the progress of the current leading player). All these characteristics suggest that 7W is a difficult challenge for AI.

In a 7W game, from 3 to 7 players[1] are first given a random personal board among the 7 available, before playing the so-called 3 ages (phases) of the game. At the beginning of each game age, each player gets a hidden hand of 7 cards. Then there are 6 rounds, where every player simultaneously selects a card from his hand and plays it, either:

- putting it on the table in his personal space, at some cost;
- or putting it under his personal board to unlock some specific power, at some cost;
- or discarding it for 3 units of the game money.

The discarding decision (or move) is always possible, while the first two possible moves depend on the player ability to gather enough resources from his board or from the production cards he already played in his personal space in order to pay the cost. He can also use game money to buy resources from cards played by his left and right neighbors. This trading decision cannot be opposed by the opponent player(s) and the price is determined by the set of cards already played.

---

[1] While the rule allows 2 player games, these are played by simulating a 3rd "dumb" player.

After playing their card, there is a so-called drafting phase, where all players give their remaining cards in hand to their left (age 1 and 3) or to their right (age 2) neighbor. This circular permutation of the cards reduces the hidden information: one can deduce some part or all of the other players game at a given point that depends on the number of player and on cards played hidden under the personal boards. When there are less than 6 players, some cards from his original hand will eventually come back to every player. On the 6th turn, when the players have only two cards remaining in their hand, they play one of the two and discard the other (except with some player board conditions).

The goal of the game is to score the maximum victory points (VP), which are awarded to the players at the end of the game, depending on the cards played on the table, under the boards and the respective amounts of game money. The cards are almost all different, but come in families distinguished by color : resources (brown and gray), military (red), civil (blue), trade (yellow), sciences (green) and guilds (purple). The green family is itself sub-divided between three symbols used for VP count.

This game presents several interesting traits for AI research, that also probably explain its success among gamers:

- it has a complex scoring scheme combining linear and non linear features: blue cards provide directly VPs to their owner, red cards offer points only to the owner of the majority of red cards symbols, yellow ones allow to save or earn game money, green ones give their owner the number of identical symbols to the square, with extra points for each pack of three different symbols.
- resource cards have delayed effect: they mainly allow a player to put VPs awarding cards on later turns; this is also the case of green cards that, apart from the scoring of symbols, allow some other cards to be played for free later on.
- there is hidden information when the players receive their hand of cards at the beginning of each age of the game.
- there is a great interactivity between players as they can buy resources from each others to achieve the playing of their own cards. Some cards also give benefits or VPs depending on which cards have been played by the neighbors. Moreover the drafting phase confronts players with the difficult choice of either playing a card that gives them an advantage, or another less rewarding card that would advantage a neighbor after the drafting phase.
- the game is strongly asymmetric relatively to the players, since all player boards are different and provide specific powers (such as resources, or military symbols). Thus some boards are oriented towards specific strategies, such as maximizing the number of military symbols, or collecting green cards symbols, for example.

The number of different cards (68 for 3 players, from those 5 are removed randomly) and the 7 specific boards, together with the delayed effect of many cards and the non linear scoring, make it difficult to handcraft an evaluation function. Notably, the number of VPs gained in the two first game ages (card deals) is a bad predictor of the final score, since scoring points at this stage of the game usually precludes playing resource cards that will be needed later on.

We can give an approximation of the state space size for 3 players, by considering that there are 68 possible different cards, from those each player will usually play 18 cards. We thus obtain $\binom{68}{18} \times \binom{50}{18} \times \binom{32}{18} = 1E38$ possible combinations, neglecting the different boards and the partition between on-table and under-the-board cards that would increase that number.

3   GP Individual Architecture

Devising a new strategy for a complex board game is typically choosing the best move (or decision) for the player at every decision step of the game. As explained in Sect. 2, a player strategy for 7W is to choose the hopefully "best" couple composed of a card and a way how to play it (on the board, under the board or discard). Our goal is to evolve a program that is able to cope well enough on average with every game position.

For this purpose, the design of an evaluation function, able to rate how a given board is promising for the current player, is a classic mean of obtaining an artificial player. Once this evaluation function has been developed, one can implement a variant of the well known minimax search algorithm. However, as explained in the presentation of the game, crafting such a function from scratch is very challenging in the case of 7W, due to the many different and often delayed ways of scoring points.

It feels natural to try and see if GP could evolve such an evaluation function. We partly proceed along the lines explored for chess endgames in [5]:

- in a similar way, we want to minimize the depth of the search, using less brute-force power and so use an efficient evaluation function for the current position;
- to the contrary, we will not use complex terminals for GP, for two motives. First there is not much expert knowledge at disposal to suggest complex GP inputs for 7W. Second, we want to try and obtain a good Koza's "A to I" ratio: can GP work out competitive players from raw game attributes ? We will see that the answer is positive.

Terminal Set. The terminal set is the most sensitive part: it should ideally embrace the whole set of variables that are accessible to predict the game evolution. Our terminal set is divided into two parts:

- a subset of 17 constant values of type real, regularly spaced in the interval $[-2, 2]$, intended to provide raw components for GP to build expressions;
- a subset of game related terminals. They try to embrace what a casual player may use among the information provided by the current game state. For example, a player will have a look at the military power of his left/right opponent, he may evaluate the number of victory points awarded by his science cards, look at how many different resources he has, and so on. As said above, we

want as much as possible to avoid complex terminals, the sole exception being Xvp that computes the total number of victory points already gained by the player (a precise computation usually needs too much time for a human to do during play).

All game related terminal are listed in Table 1. Their computational complexity is low, so they will guarantee a quick GP player.

Table 1. Game related terminal set for "Seven Wonders"

Terminal	Value type	Note
Xmil	int	military strength of the current player
Xlmil	int	military strength of the left player
Xrmil	int	military strength of the right player
Xtrade	int	number of own commercial cards (yellow cards)
Xtradrsc	int	number of own yellow cards providing resources
Xcivil	int	number of own civil cards
Xrsc	int	number of own resource cards
Xdiffrsc	int	number of own different resource types
Xscience	int	number of own science cards
Xsciencep	int	victory points earned by own science cards
Xgold	int	number of own gold coins
Xchain	int	number of cards that could be chained (i.e. played for free)
Xage	int	current age (i.e. card deal number)
Xturn	int	current ply of the current age
Xvp	int	total victory points earned so far by the player

Many more terminals, reflecting various information that a player can use, could be added. For instance most of the terminals introduced in Table 1 can be derived with a left and right version to check the same information for opponents, such as is already done with the military strength. This could also be extended to cover non-neighbors players (even if interacting with them is somewhat limited by the game rules). Other information that could be provided is the player next round hands of cards, that can be either simulated or known exactly (due to the circular permutation of cards at each game round).

Function Set. The function set is kept simple. Following the classic example of Shannon's evaluation function for a chessboard [7] that computes a linear weighted sum of the number of pieces and of some piece combinations, we decided to restrict the function set to the 3 basic operations $\{+, -, *\}$. Note that even if the set may seem limited, it allows non linear combinations of the terminals.

4   Computing Individual Fitness

The fitness evaluation is a tricky part when we have to evaluate a player. It is quite obvious that playing against an under-average player or a random player will not provide interesting feedback. Furthermore as the game has some stochastic features, a player may win or lose even if his average level is better than those of his opponents. This means that we must be very careful when evaluating the GP player.

In order to evaluate the GP individuals, we need to oppose them to other AIs. In the absence of any human designed evaluation function that could be used in a minimax framework for providing an opponent, we opted for the development of a rule-based AI, and a MCTS player, that are presented below. Another solution could have been to oppose a GP player to another (or some others) from the GP population: first experiments seemed less promising, so this solution was put aside for the moment.

4.1   A Rule-Based AI

Designing rules for 7W seems rather easy and feels close to the way a beginner player can proceed, considering the cards already played, those in hand, and deciding one's next move. Our rule-based AI (rule-AI) follows a set of rules of decreasing priority, stated below, for the two first deals of a game (when a card is said as being "always played", it means of course if its cost is affordable):

- a card providing two or more different resource types is always played;
- a card providing a single resource type that is lacking to the rule-AI is always played;
- a military card is always played if rule-AI is not currently the only leader in military strength, and the card allows rule-AI to become the (or one of the) leading military player(s);
- at random either the civil card with the greatest victory points (VP) award or one science card is always played;
- a random remaining card is played;
- as a default choice, a random card is discarded.

Contrasting with the first two deals that mainly involve investment decisions, the last deal of the game, in the so-called third age, appears as the time to exploit the previous choices: the set of rules is now replaced by choosing the decision with best immediate VP reward.

Clearly we do not hope to reach championship level with such a simplistic set of rules, nonetheless the resulting player is able to beat human beginners. A comparison with MCTS success rate is given below in Sect. 4.3.

4.2   MCTS Player

The Monte-Carlo Tree Search algorithm (MCTS) has been recently proposed for decision-making problems [4,8,9]. Applications are numerous and varied, and encompass notably games [10–14]. In games when evaluation functions are hard to design, MCTS outperforms alpha-beta techniques and is becoming a standard approach. Probably the best known implementation of MCTS is Upper Confidence Tree (UCT), presented below.

The idea is to build an imbalanced partial subtree by performing many random simulations from the current state of the game, and simulation after simulation biasing these simulations toward those that give good results, thus exploring the most promising part of the game tree. The construction is done incrementally and can be described as consisting in four different parts: *descent* in the partial subtree, *growth* by adding a new child under the current leaf node, *evaluation* of the current branch by a random simulation of the end of the game, and *update* of the current branch information.

The *descent* is done by using a bandit formula, i.e. at node s, among all possible children C_s, we choose to descend on next node $s' \in C_s$ that gives the best reward according to the formula:

$$s' \leftarrow \arg\max_{j \in C_s} \left[\bar{x}_j + K_{UCT} \sqrt{\frac{ln(n_s)}{n_j}} \right]$$

with $\bar{x}_j$ the average reward for the node j (it is the ratio of the number of victories over the number of simulations), n_j the number of simulations for node j and n_s is the number of simulation for the node s, with $n_s = \sum_j n_j$. The constant K_{UCT} is an exploration parameter used to tune the trade-off between exploitation and exploration. At the end of the *descent* part, a node which is outside the subtree has been reached and is added to the sub-tree (unless this branch already reach the end of the game). In order to *evaluate* this new node, a so-called *playout* is done: random decisions are taken until the end of the game, when the winner is known and the success ratio of the new node and all its ascendant are *updated*.

We refer the reader to the literature cited at the beginning of this subsection for more information about MCTS, and also to [15] for a more detailed presentation of our MCTS dedicated to 7W, notably how to handle partial information via determinization (see also [16] on this topic): we simply sketch some implementation details in the following. A single N-player game turn (corresponding to the N player simultaneous decisions), is represented in the MCTS subtree by N successive levels, thus for a typical 3-player game with 3 ages and 6 cards to play per age, we get $3 \times 6 = 18$ decisions per player and the depth of the tree is $18 \times 3 = 54$. Of course we keep the simultaneous decisions, that is the state of the game is updated only when reaching a subtree level whose depth is a multiple of N, thus successive players (either real or simulated) make their decision without knowing their opponent choices for the curent turn. The average node arity can be estimated empirically at an average of 14 children per node, and a good value for K_{UCT}, also obtained empirically is between 0.3 and 0.5.

To implement MCTS one just need to know how to generate the possible decisions for the current state of the game. Its drawback is its running time: obtaining a good level of play typically implies to simulate the completion (playout) of several thousands games. To obtain a better trade-off between speed and quality, we increase the number of playouts at game age 2 and 3 since the completion of the game simulations is shorter: when we state N simulations, we mean N for the second game deal, $0.66 * N$ for the first deal, and $1.33 * N$ when playing the last deal (thus it is N simulations on average for the whole game).

The tuning of MCTS, notably the K_{UCT} constant, can be a bit tricky: we presented this in [15], with the comparison of several MCTS enhancements (such as incorporating the game score in the bandit formula). In the next sections we use a refined value for $K_{UCT} = 0.5$, that appears slightly better, and we present some new results.

4.3   Comparison of Rule-AI and MCTS

To serve as a reference in Sect. 5, we compared the success rates of the two AIs previously described. When we opposed two rule-AIs to the MCTS player with 1500 and 3000 playouts and $K_{UCT} = 0.5$ we obtain the resulting success rates on 5000 games:

As expected the two clones of rule-AI obtains very similar success rates. The MCTS is a better player than rule-AI, and the bigger the number of simulations allowed per move for MCTS, the better the success rate, as expected in theory.

4.4   Choosing Moves and Assigning Fitness for GP

In order for the GP individual to choose a move, we examine the resulting game state of all possible GP moves, together with a random move of the opponents (remember that all players move simultaneously). We usually do not know exactly which cards are in the opponent hands, but we maintain the set of the possible cards in order to sample their possible decisions, doing N *determinizations*. That means that we simulate a random choice of opponents moves on their potential cards N times for each decision of the GP player (see e.g. [15,16] for an illustration of this technique). We select the GP move that is associated to the biggest evaluation function value summed (or equivalently averaged) on the set of determinizations. This is indeed equivalent to an expectimax search of depth one.

Ideally an expectimax search should sample every possible opponent moves. In our case, as there can be up to 1764 possible combinations of opponent moves due to incomplete information, this would not be practically feasible: remember this must be done for all moves, in all games, for all individuals and all generations. Thus we fix the number of determinizations to $N = 11$, for the sake of rapidity: this is a low value but it already yields a satisfying level of play.

When GP needs the fitness of an individual, we have to assess its quality as an evaluation function. We proceed by playing a set of P games where the GP individual is opposed to other AIs. Finally the fitness is the success ratio (percentage of the games won) obtained by GP on the P games. It proved necessary to train GP on several hundred games to obtain a suitable fitness. Early experiments on 100 independent games gave fragile GP players, that lost almost always with some configurations of player boards (there are 210 such configurations). We settled on 500 games, composed with 25 different random player board configurations and 20 deals per configuration, to assess one individual fitness. Again for the sake of speed, we chose ruleAI as the only opponent for training GP since it is much faster than MCTS, but MCTS could still be used for validation.

Table 2. Comparisons of success rates (SR) for two identical rule-AIs and one MCTS player with either 1500 or 3000 simulations per move.

# MCTS simulations	rule-AI-0	rule-AI-1	MCTS
1500	18.92 % ± 1.09	21.10 % ± 1.13	59.98 % ± 1.36
3000	17.80 % ± 1.06	17.76 % ± 1.06	64.44 % ± 1.33

5   Experiments and Results

We recall that we evolve GP players (strictly speaking evaluation functions) trained against ruleAI, described previously, and the fitness is the success ratio obtained on 500 games. Each GP move is chosen as the one bringing the greatest evaluation value by a depth one expectimax on 11 random determizations of opponent moves.

The default GP parameters are: population size of 25 individuals only, tournaments of size 2, "steady state" worst individual replacement, crossover probability 0.6, point mutation probability 0.15 and 100 generations. The population size is small compared to usual practice, but it was required to reduce the computing time: a GP run against rule-AI still takes two days on a 32 cores Intel Xeon CPU E5-4603 v2 @ 2.20 GHz machine with multi-threaded fitness computation (the parallelization is done on the 500 games needed to assess a significant fitness). We chose a crossover probability rather less than standard, with a higher than usual mutation rate, for the sake of keeping more diversity in such a small population. Time constraints prevented us to perform a systematic study of these parameters, however the results derived from this setting are already satisfying.

Once a GP player is evolved, its success ratio is validated on 5000 games. Half of the validation games are played on the same 25 board configurations used for training, the other half is played on random configurations, and in both cases the card deals are completely independent from the learning phase. Note that while evolution is time consuming, the resulting GP program plays almost instantly. Validation on 5000 games against rule-AI takes less than an hour, while against MCTS it still takes several days, due to the MCTS computing cost.

5.1   Training GP versus 2 Rule-AIs

When trained against two clones of rule-AI, GP yielded a best individual with fitness 0.754 at generation 93, that is the evolved player wins three games over four, while the expected win rate is 0.33 if it were of the same strength as its opponents. Once simplified (with the SAGE symbolic calculus package) this best individual, listed in Fig. 1, is amenable to some analysis and interpretation.

```
(6*Xdiffrsc + 8*Xmil + 5*Xtrade + 9.5*Xtradrsc + 5*Xvp
- 4.875)*Xtradrsc + 4*Xdiffrsc + 5*Xmil + 5*Xsciencep +
3.0*Xtrade + 17*Xtradrsc + Xvp - 11.5
```

Fig. 1. Best individual trained against 2 rule-AIs, winning 70 % of games (obtained at generation 90, average fitness 0.56)

Table 3. Comparisons of success rates (SR) for two identical rule-AIs and the GP player of Fig. 1.

rule-AI-0	rule-AI-1	GP
22.20 % ± 1.15	21.52 % ± 1.14	56.28 % ± 1.37

While this individual is not a linear function, it remains a rather simple quadratic polynomial and still looks like a traditional weighted combination of attributes. We can see that some terminals are associated to strong weights, notably Xtradsrc the number of cards providing a choice of alternative resources, Xmil the military strength and Xdiffrsc the number of different available resources. Intuitively these are important inputs since having access to different resources help in developing one's game and it is rather difficult to win a game without any success in the military strategy. Indeed our rule-AI uses similar information as highest criteria for decision taking — but with much less success!

The validation experiment is presented in Table 3. The GP fitness being measured on the training set, it proved to be too optimistic, as expected in theory. The validation win rate is nonetheless superior to rule-AI and unexpectedly very close to the success rate of MCTS parametered with 1500 simulations per move (reported in Table 2). It feels rather remarkable that GP could evolved such a successful formula, that is:

- unique for the whole game,
- using only raw game attributes,
- able to choose the next move with a search of depth only one, which is almost instantaneous.

5.2   Validation Against MCTS with 1500 and 3000 Playouts

In Table 4 we validate our best GP individual against rule-AI and MCTS. GP is the best of the three, while MCTS wins the second rank. It seems curious since MCTS scored better than GP when opposed only to ruleAI, but this is an illustration of the difficulties raised when opposing three players: the strategies of two players may combined to the detriment of the last. One can notice than

Table 4. Comparisons of success rates (SR) for one rule-AI, one MCTS with 1500 playouts, and the GP player of Fig. 1.

# MCTS simulations	rule-AI	MCTS	GP
1500	24.14 % ± 1.22	34.13 % ± 1.35	41.72 % ± 1.40
3000	20.40 % ± 1.12	39.11 % ± 1.35	40.49 % ± 1.36

MCTS still wins at least one third of the games so it means that it is rule-AI which gives way to either GP or MCTS. By contrast, GP appears rather robust in this context against the increase in MCTS simulations.

In Table 5, MCTS is opposed to 2 clones of GP. With 1500 playouts MCTS is no more able to win one third of the games, thus clearly meaning that GP has a superior play ability. But once the number of playouts is raised to 3000, this time MCTS is the winner. Again, it is an illustration of the dependencies between more than two players: there is no weak rule-AI player that GP can loot, and probably the two similar GP individuals hinder themselves by using the same strategy.

Table 5. Comparisons of success rates (SR) for one MCTS with 1500 playouts, against two GP player clones of Fig. 1.

# MCTS simulations	GP-1	GP-2	MCTS
1500	34.40 % ± 1.32	35.34 % ± 1.33	30.26 % ± 1.27
3000	32.34 % ± 1.30	31.82 % ± 1.29	35.84 % ± 1.33

These results show that GP can evolve successful players, competitive against MCTS, the current method of choice when evaluation functions are not easy to obtain. Notice that even if our GP individual plays remarkably against the other artificial opponents, it is not yet tough enough to deal with experienced human players. The absence of information about opponent moves is a strong limitation that could be exploited by humans.

On the one hand MCTS keeps the advantage of being improved simply by increasing the number of simulations (although it may become too slow to be acceptable), while on the other hand the GP player is several order of magnitude faster but cannot be improved as easily.

6 Conclusion and Future Works

This study has shown that GP can evolve very competitive players for complex board games in the eurogame category, even from basic inputs. The main technical problem we encountered was the huge amount of computing time needed to obtain a significant fitness. This prevented us, at least for the moment, to obtain GP players trained against MCTS. We stress again that, on the opposite, once evolved, the resulting player is almost instantaneous, several orders or magnitude faster than MCTS.

Training only against the ruleAI, which is a weak player, nonetheless allowed GP to beat MCTS with 1500 playouts on average, and to compete with a 3000 playout MCTS in a mixed players context. This a significant result which was unexpected, especially as 3000 playouts already incurs a significant delay for MCTS.

Many tracks are opened by this study. Some are GP oriented such as co-evolving programs by assessing fitness against the other individuals of the population; or trying smarter terminals and expanding the function set to include e.g. "if" statements; or splitting the program in three subroutines, one for each deals of the game, in order to obtain an increased level of play by adjusting the evaluation function to the current phase of the game. Tuning the evolution parameters, and also the number of determinizations are natural extensions, and also testing our GP evaluation function in an expectimax of depth greater than one.

A more fundamental idea could be to try to bridge the gap between GP and MCTS, e.g. using GP as a surrogate estimation of a fraction of the playouts, in order to speed up MCTS, or using GP to build hyper-heuristics using MCTS sampling, GP evolved evaluation functions and even rule-AI as building bricks. We could also try to learn game patterns, following the ideas in [17]. Other possible MCTS variants could use our ruleAI or GP player to bias the MCTS sampling in playouts. At last, tackling other eurogames is also a future objective.

References

1. Pfeiffer, M.: Reinforcement learning of strategies for Settlers of Catan. In: Mehdi, Q., Gough, N., Natkin, S., Al-Dabass, D. (eds.) 5th international conference on computer games: artificial intelligence, design and education, pp. 384–388 (2004)
2. Szita, I., Chaslot, G., Spronck, P.: Monte-Carlo tree search in settlers of catan. In: van den Herik, H.J., Spronck, P. (eds.) ACG 2009. LNCS, vol. 6048, pp. 21–32. Springer, Heidelberg (2010)
3. Winder, R.K.: Methods for approximating value functions for the Dominion card game. Evol. Intell. 6(4), 195–204 (2013)
4. Chaslot, G.M.J.P., Saito, J.-T., Bouzy, B., Uiterwijk, J.W.H.M., Van Den Herik, H.J.: Monte-Carlo strategies for computer Go. In: Proceedings of the 18th BeNeLux Conference on Artificial Intelligence, Namur, Belgium, pp. 83–91 (2006)
5. Hauptman, A., Sipper, M.: GP-endchess: using genetic programming to evolve chess endgame players. In: Keijzer, M., Tettamanzi, A.G.B., Collet, P., van Hemert, J., Tomassini, M. (eds.) EuroGP 2005. LNCS, vol. 3447, pp. 120–131. Springer, Heidelberg (2005)
6. Sipper, M.: Evolving game-playing strategies with genetic programming. ERCIM News 64, 28–29 (2008). Invited article
7. Shannon, C.E.: XXII. Programming a computer for playing chess. Philos. Mag. (Ser. 7) 41(314), 256–275 (1950)
8. Kocsis, L., Szepesvári, C.: Bandit based Monte-Carlo planning. In: Fürnkranz, J., Scheffer, T., Spiliopoulou, M. (eds.) ECML 2006. LNCS (LNAI), vol. 4212, pp. 282–293. Springer, Heidelberg (2006)

9. Coulom, R.: Efficient selectivity and backup operators in monte-carlo tree search. In: Herik, H.J., Ciancarini, P., Donkers, H.H.L.M.J. (eds.) CG 2006. LNCS, vol. 4630, pp. 72–83. Springer, Heidelberg (2007)
10. Gelly, S., Silver, D.: Combining online and offline knowledge in uct. In: Proceedings of the 24th International Conference on Machine learning, pp. 273–280. ACM (2007)
11. Lorentz, R.J.: Amazons discover Monte-Carlo. In: Herik, H.J., Xu, X., Ma, Z., Winands, M.H.M. (eds.) CG 2008. LNCS, vol. 5131, pp. 13–24. Springer, Heidelberg (2008)
12. Cazenave, T.: Monte-Carlo kakuro. In: van den Herik, H.J., Spronck, P. (eds.) ACG 2009. LNCS, vol. 6048, pp. 45–54. Springer, Heidelberg (2010)
13. Arneson, B., Hayward, R.B., Henderson, P.: Monte-Carlo tree search in Hex. IEEE Trans. Comput. Intell. AI Games 2(4), 251–258 (2010)
14. Teytaud, F., Teytaud, O.: Creating an upper-confidence-tree program for havannah. In: van den Herik, H.J., Spronck, P. (eds.) ACG 2009. LNCS, vol. 6048, pp. 65–74. Springer, Heidelberg (2010)
15. Robilliard, D., Fonlupt, C., Teytaud, F.: Monte-Carlo tree search for the game of "7 wonders". In: Cazenave, T., Winands, M.H.M., Björnsson, Y. (eds.) CGW 2014. CCIS, vol. 504, pp. 64–77. Springer, Heidelberg (2014)
16. Whitehouse, D., Powley, E.J., Cowling, P.I.: Determinization and information set Monte-Carlo tree search for the card game Dou Di Zhu. In: IEEE Conference on Computational Intelligence and Games (CIG), pp. 87–94. IEEE (2011)
17. Hoock, J.-B., Teytaud, O.: Bandit-based genetic programming. In: Esparcia-Alcázar, A.I., Ekárt, A., Silva, S., Dignum, S., Uyar, A.Ş. (eds.) EuroGP 2010. LNCS, vol. 6021, pp. 268–277. Springer, Heidelberg (2010)

SGE: A Structured Representation for Grammatical Evolution

Nuno Lourenço[1]([✉]), Francisco B. Pereira[1,2], and Ernesto Costa[1]

[1] CISUC, Department of Informatics Engineering, University of Coimbra,
Polo II - Pinhal de Marrocos, 3030-290 Coimbra, Portugal
{naml,xico,ernesto}@dei.uc.pt
[2] Polytechnic Institute of Coimbra, Quinta da Nora, 3030-199 Coimbra, Portugal

Abstract. This paper introduces Structured Grammatical Evolution, a new genotypic representation for Grammatical Evolution, where each gene is explicitly linked to a non-terminal of the grammar being used. This one-to-one correspondence ensures that the modification of a gene does not affect the derivation options of other non-terminals, thereby increasing locality. The performance of the new representation is accessed on a set of benchmark problems. The results obtained confirm the effectiveness of the proposed approach, as it is able to outperform standard grammatical evolution on all selected optimization problems.

1 Introduction

Evolutionary Algorithms (EA) are computational methods inspired by the principles of natural selection and genetics. Over the years they have been successfully used in different situations, including optimization, design or learning problems. Genetic Programming (GP) is an EA branch that is able to automatically evolve computer programs/algorithmic strategies. One of the most relevant variants of GP is Grammatical Evolution (GE), whose distinctive feature is how it decouples the genotype (a linear string) from the phenotype (a tree expression). GE relies on a mapping process to translate the linear string into an executable program. This transformation is guided by grammar production rules that help to establish the set of syntactically correct programs.

The aim of this paper is to propose Structured Grammatical Evolution (SGE), an enhanced genotypic representation for GE. In SGE there is a one-to-one mapping between genes and non-terminals belonging to the grammar. In order to allow a valid mapping, each gene encodes a list of integers that represent the possible derivation choices of the corresponding non-terminal. The structured representation of SGE, in which a gene is explicitly linked to a non-terminal, ensures that changes in a single genotypic position do not affect the derivation options of other non-terminals. By removing these interactions, SGE might help to solve some well-known locality issues that affect GE [9]. In the next sections we describe the application of SGE to several GP benchmarks problems [12] and compare its performance against a standard GE approach. The optimization results confirm the effectiveness and efficiency of SGE.

© Springer International Publishing Switzerland 2016
S. Bonnevay et al. (Eds.): EA 2015, LNCS 9554, pp. 136–148, 2016.
DOI: 10.1007/978-3-319-31471-6_11

The remainder of the paper is organized as follows: Sect. 2 provides a brief introduction to GE and reviews relevant contributions dealing with GE representation. Section 3 introduces SGE and details the genotype-phenotype mapping, whereas Sect. 4 comprises the optimization study. Finally, Sect. 5 gathers the main conclusions and presents some ideas for future work.

2   Grammatical Evolution

Grammatical Evolution (GE) is a form of Grammar-Based Genetic Programming (GBGP) [6]. As with standard GP, the goal of GE is to evolve executable algorithmic strategies. GE is different from other non grammar-based GP variants, for there is a separation of the genotype, a linear string, and the phenotype, a program in the form of a tree expression. As a consequence, a mapping process is required to map the string into an executable program, using the productions rules of a context-free grammar (CFG). A CFG is a tuple $G = (N, T, S, P)$, where N is a non-empty set of non-terminal symbols, T is a non-empty set of terminal symbols, S is an element of N called axiom, and P is a set of production rules of the form $A ::= \alpha$, with $A \in N$ and $\alpha \in (N \cup T)^*$. N and T are disjoint. Each grammar G defines a language $L(G)$ composed by all sequences of terminal symbols (the words) that can be derived from the axiom: $L(G) = \{w : S \overset{*}{\Rightarrow} w, w \in T^*\}$.

The translation of the genotype into the phenotype is done by simulating a leftmost derivation from the axiom of the grammar. This process scans the linear sequence from left to right and each integer (*i.e.*, each codon) is used to determine the grammar rule that expands the leftmost non-terminal symbol of the current partial derivation tree. Suppose that we have the following production rule,

$$
\begin{aligned}
< expr > ::= &\ < expr >< op >< expr > & (0)\\
| &(< expr >) & (1)\\
| &< pre - op > (< expr >) & (2)\\
| &< var > & (3)
\end{aligned}
$$

where there are four options to rewrite the left-hand side symbol $< expr >$. In the beginning we have a sentential form equal to the axiom $< expr >$. To rewrite the axiom one must choose which alternative will be used by taking the first codon and dividing it by the number of options for $< expr >$. The remainder of that operation will indicate the option to be used. In the example above, assuming that the first integer is 8, it follows that $8 \% 4 = 0$ and the axiom is rewritten in $< expr >< op >< expr >$. Then the second integer is read, and the same method is used to the left most non-terminal of the derivation. Sometimes the length of the string is not sufficient to complete the mapping. In those cases the sequence is repeatedly reused in a process known as wrapping. If mapping exceeds a pre-determined number of wrappings, the process stops and the worst possible fitness value is assigned to the individual.

2.1   Other GE Representations

There are some reports in the literature describing enhancements to the standard GE representation and mapping. The *bucket rule* from Keijzer et al. [3] allows a given codon value to select different production choices, thereby removing the bias created by the order of the grammar entries.

In [7], O'Neill et al. presented the Position Independent GE (πGE), an alternative genotype-phenotype mapping. In the traditional GE mapping there is a positional dependency, as the derivation is always performed by expanding the leftmost terminal in the derivation tree. πGE removes this dependency by creating codons with two values: *nont* and *rule*. In this case, *nont* helps to select the next non-terminal NT to be expanded: $NT = nont\%count$, where *nont* is the value present in the genotype, and *count* is the number of non-terminals still in the derivation tree. The *rule* value of the codon pair, as in standard GE, selects which production rule should be applied from the selected non-terminal NT.

Chorus [10] is an alternative proposal aiming at developing a position independent GE, although the results presented in the above mentioned reference do not show any relevant advantage over standard GE.

Fagan and coworkers [1] compared the performance of several mapping mechanisms. Besides the aforementioned πGE and the traditional depth-first expansion they considered two additional methods, breadth-first and a random expansion mechanism, and concluded that πGE provides advantages over standard GE. This result confirms that it is worthwhile to investigate new, alternative, genotypic representations, together with the mapping process.

3   Structured Grammatical Evolution

In SGE each gene is linked to a specific non-terminal and is composed by a list of integers. The length of each list is determined by computing the maximum possible number of expansions of the corresponding non-terminal (see details in Sect. 3.1). This structure ensures that when a gene is modified, it does not affect the derivation options of other non-terminals, thus narrowing the number of changes that occur at the phenotypic level.

The values that are inside the lists correspond to the number of possible expansion choices. Therefore, when performing the mapping it is possible to remove the modulo rule, thus reducing the redundancy associated with it. Consider the following set of production rules:

$$< start >::= < int > \mid < int > * < int >$$
$$< int >::= 1|2|3|4$$

There are two non-terminals $\{< start >, < int >\}$. The genotype is composed by two genes, where the first gene is linked to $< start >$, and the second to $< int >$. Then it is necessary to compute the length of the gene's lists by calculating the maximum number of expansions of a non-terminal. The $< start >$ symbol is expanded only once, as it is the grammar axiom. The $< int >$ symbol is expanded, at most, twice, because of the rule $< int > * < int >$. Thus the lists will have length 1 and 2, respectively. Finally, to fill them we count the number of possible derivation options, c_N, of each non-terminal and assign to each position of the list a random value from the interval $[0, c_N - 1]$. Considering the example above, the $< start >$ symbol has $c_N = 2$ and $< int >$ has $c_N = 4$. Two possible genotypes are depicted in Fig. 1.

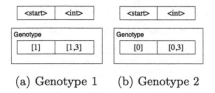

(a) Genotype 1   (b) Genotype 2

Fig. 1. SGE: example of two possible genotypes

The process of translating a genotype into a phenotype is similar to the standard GE mapping. This process starts by expanding the axiom of the grammar, and then expanding the non-terminals in a left-first manner. Consider the example above, where the axiom is the non-terminal $< start >$. To expand it, we look into its gene within the genotype (Fig. 1a). The first unused integer of the list is 1, which selects the option $< int > * < int >$. The next symbol to be rewritten is $< int >$. Its first unused integer is 1, thus it is replaced by the option "2". Next the second $< int >$ is expanded. The first unused integer in the associated gene is 3, which dictates the option "4" should be selected. As there are no more symbols to expand, the process ends, and returns the phenotype: "2*4". The phenotype associated with the genotype of Fig. 1b is "1".

3.1 Pre-processing

The first step to construct the genotype is to compute an upper bound for the number of times that a non-terminal can be expanded as it defines the list size for each gene. Initially, we iterate through the productions belonging to the grammar, and record the maximum number of references to non-terminals that occur in each choice (Algorithm 1). At the same time we build a set that dictates a relation between non-terminals.

Finally, we iterate the set of non-terminals and determine recursively the number of times that, at most, each non-terminal will be expanded (Algorithm 2).

Consider the following set of production rules, with $< start >$ as the axiom:

$$< start > ::= < line > \mid < line > / < line >$$
$$< line > ::= < var > * < var >$$
$$< var > ::= x1 \mid x2 \mid 1$$

Algorithm 1. Computation of the references that exist in the grammar.

```
countReferences ← {}
isReferencedBy ← {}
for nt in nonTerminalsSet do
    for production in grammar[nt] do
        for option in production do
            if option ∈ nonTerminalsSet then
                isReferencedBy[option] ← nt
                count[option] ← count[option] + 1
            end if
        end for
    end for
    for key in count do
        countReferences[key][nt] ← max(countReferences[key][nt], count[key])
    end for
end for
```

Algorithm 2. Calculate the upper bound for the number of times that a non-terminal can be expanded.

```
function FINDREFERENCES(nt, isRefBy, countRefProd)
    r ← getTotalReferencesOfProd(countRefProd, nt)
    results ← []
    if nt = startSymbol then
        return 1
    end if
    for ref in isRefBy[nt] do
        result.add(FINDREFERENCES(ref,isRefBy,countRefProd))
    end for
    references ← references * max(result)
    return references
end function
```

Using the algorithm described above to compute the size of each gene, we obtain: $< start >$: 1, $< line >$: 2, $< var >$: 4. Then we determine the values of c_N, i.e., the number of derivation choices, for each non-terminal: $< start >$: 2, $< line >$: 1, $< var >$: 3.

3.2   Recursive Grammars

The pre-processing described in the previous section does not consider recursive grammars. Standard GE deals with recursion by always trying to perform the translation into an executable program. If it runs out of integers, GE assigns the worst possible fitness value to the individual.

SGE deals with recursion in a different way, as it follows a preemptive approach: a maximum level of recursion must be defined beforehand. Hence it is necessary to introduce a set of intermediate symbols that mimic the levels of the recursion tree. The following example is an excerpt of a grammar for symbolic regression problems:

$$< start > ::= < expr >$$
$$< expr > ::= < expr >< op >< expr > \mid < var >$$

$$< op >::= + \,|\, - \,|\, * \,|/$$
$$< var >::= x$$

Looking into the grammar, we see that the $< expr >$ production is recursive. Therefore it needs to be rewritten. Assuming that 2 levels of recursion were defined it becomes:

$$< start >:= < expr >$$
$$< expr >::= < expr\_lvl\_0 >< op >< expr\_lvl\_0 >$$
$$|\, < var >$$
$$< expr\_lvl\_0 >::= < expr\_lvl\_1 >< op >< expr\_lvl\_1 >$$
$$|\, < var >$$
$$< expr\_lvl\_1 >::= < var >< op >< var > \,|\, < var >$$
$$< op >::= + \,|\, - \,|\, * \,|/$$
$$< var >::= x$$

While transforming the grammar we ensure two things: first, that all the symbols have the same probability of being selected after the transformation, because they are copied to each new added level; second, that there will be no invalid individuals, since the mapping process always ends.

All GP variants impose a constraint in the maximum program size, a mandatory step to prevent solutions from growing excessively and becoming computationally intractable. The constraint might be imposed in terms of tree depth, number of available nodes [4], or by imposing limits on the number of wrappings as performed in GE [6]. Following a similar line of procedure, SGE limits the maximum program size by imposing a limit on the number of recursive calls.

3.3   Genetic Operators

GE relies on standard operators to navigate the search space looking for promising solutions to the problem at hand. Two existing variation operators are adapted to work with SGE.

Recombination. This operator is an adaptation of the uniform crossover for binary representations. It starts by creating a binary mask with the same length of the genotype. Then the offspring are created by selecting the parents genes based on the mask values. Recombination does not modify the values of the lists inside the genes. Figure 2 illustrates an application of this operator.

Mutation. This operator is based on the integer flip mutation. A gene is mutated by randomly selecting a position inside the list and changing it to a new random value from $[0, c_N - 1]$.

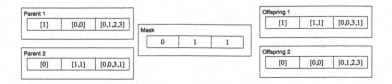

Fig. 2. Application of the recombination operator

4   Experimental Analysis

To validate SGE, three problems were chosen following the guidelines proposed by White et al. to select good GP benchmarks [12]: harmonic curve regression, polynomial regression, and the Santa Fe Ant trail.

4.1   Problems Description

Harmonic Curve Regression. The goal is to approximate the series defined by

$$\sum_i^x \frac{1}{i} \tag{1}$$

where $x \in [1, 50]$. This problem is interesting as it complements the standard interpolation task with a generalisation step. In this second stage, the interval $x \in [51, 120]$ is considered. The production set for the harmonic curve regression is defined as:

$$< start > ::= < expr >$$
$$< expr > ::= < expr > < op > < expr > | (< expr >)$$
$$| < pre\_op > (< expr >) | < var >$$
$$< op > ::= + | *$$
$$< pre\_op > ::= + | - | inverse | sqrt$$
$$< var > ::= x$$

where *inverse* is $1/x$.

Pagie Polynomial. This is a hard symbolic regression problem [12], where the goal is to approximate the polynomial function defined by:

$$\frac{1}{1 + x^{-4}} + \frac{1}{1 + y^{-4}} \tag{2}$$

The function is sampled over the range $[-5, 5]$, with a step $s = 0.4$. The production set for this problem is defined as:

$$< start > ::= < expr >$$

$$< expr > ::= < expr >< op >< expr >$$
$$|(< expr >)$$
$$| < pre\_op > (< expr >)$$
$$| < var >$$
$$< op > ::= + | - | * | /$$
$$< pre\_op > ::= sin|cos|exp|log$$
$$< var > ::= x|y$$

Artificial Ant. The goal is to evolve a strategy that an agent will follow to collect 89 pieces of food along the Santa Fe Ant trail. The production set used is the same as in [6].

$$< start > ::= < code >$$
$$< code > ::= < line >$$
$$| < code >$$
$$< line >$$
$$< line > ::= if\ ant.sense\_food() :$$
$$< line >$$
$$else :$$
$$< line >$$
$$| < op >$$
$$< op > ::= ant.turn\_left()$$
$$|ant.turn\_right()$$
$$|ant.move\_forward()$$

The maximum number of steps that the ant has to collect all the food pellets is 650 [11]. The fitness function used corresponds to the difference between the total number of food pieces available, and the number of pieces of that the ant has eaten, i.e., $(89 - \#FoodPiecesEaten)$.

4.2   Parameters

The GEVA implementation of GE was selected as the baseline of comparison for our experiments. It is an open-source implementation of Grammatical Evolution, in JAVA, and is developed and maintained by O'Neill et al. [8]. SGE was built over the GEVA search engine. There are, however, some slight changes, such as the set of variation operators used and the definition of a maximum level of recursion. The parameters for both SGE and GEVA are defined in Table 1.

We performed 30 independent runs of each approach in the optimization scenarios selected. When comparing SGE with GE a statistical analysis was done to assess if there were differences in the means and, if that was the case, how relevant they were. Since the samples do not follow a normal distribution, the analysis was performed using non-parametric tests. Moreover, and since we are dealing with two unrelated groups, the Mann-Whitney test, at a $\alpha = 0.05$ level of significance, was selected. When differences exist we compute the effect size r [2], to determine how large the differences are. For clarity, we used the following notation: a $+++$ sign indicates that the effect size is large ($r >= 0.5$), a $++$ sign indicates that the effect size is medium ($0.3 <= r < 0.5$), whereas a $+$ identifies a small effect size ($0.1 <= r < 0.3$).

Table 1. Settings for the experimental analysis

Parameter	GEVA	SGE
Initial Population	500	
Recombination rate	0.9	
Mutation rate	0.02	
Replacement	Steady-State with a generation gap of 0.9	
Selection	Tournament with size 3	
Generations	50	
Recombination Operator	Single Point Crossover	SGE Uniform Crossover
Mutation Operator	Integer Flip Mutation	SGE Integer Flip Mutation
Genotype Size	128 (Ramped Half and Half Initialization)	-
Wraps	3	-
Maximum Level of Recursion	-	6

4.3   Results

For the Harmonic Curve Regression, Fig. 3 shows the evolution of the Mean Best Fitness (MBF). An inspection of results shows that the individuals in the initial population of GE have a slightly better fitness, due to the sensible initialization method. The figure also reveals that both GE variants gradually discover better approximations as the run progresses. However SGE exhibits an increased effectiveness, rapidly discovering solutions that surpass the ones found by GE. After 12 generations SGE has already found solutions better than the overall bests of GE.

To estimate the generalization ability, we selected, for each variant (GE and SGE), the best strategy from the initial, middle (gen. 25) and final generations. We then applied the 6 selected strategies to the extended interval from the harmonic curve regression problem. The obtained errors are displayed in Fig. 4. The bars reveal that strategies discovered in later GE and SGE generations tend to obtain better results, suggesting that overfitting did not occur in the interpolation stage. Also, pairs of strategies taken in the same generation (from GE and SGE) obtain comparable results. There are never statistical significant differences, suggesting that, in this particular problem, SGE and GE have similar generalization ability. Finally, it is worth noting that the solutions evolved by SGE seem to be more reliable, as they have a global smaller standard deviation (0.24 vs. 0.4).

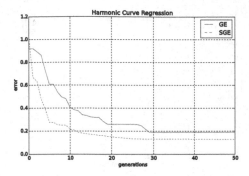

Fig. 3. Mean Best Fitness plots for the Harmonic number

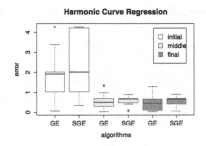

Fig. 4. Mean Best Fitness plots for the harmonic curve regression in the generalization task.

The next problem is the Pagie polynomial. The optimization results follow a trend similar to the one identified in the first problem (Fig. 5). The individuals of the initial population of SGE and GE have comparable fitness. Then, as optimization advances, SGE gradually and consistently obtains low error solutions without stagnating. On the contrary, GE exhibits a slower evolution rate and it stalls at some generations. Looking at the quality obtained by the two variants in the end of the evolutionary run, there is a noticeable difference between SGE and GE. SGE obtained solutions with considerable low error, which reinforces its effectiveness when compared with GE.

Figure 6 clearly shows that SGE outperforms GE in the Santa Fe Ant trail, the last selected benchmark. Although the initial solutions of GE have a better quality, at the end of the evolutionary process SGE provides consistently better results. This is so that in all runs, SGE was able to find solutions that allow the ant to eat all the food pieces in the board, leading to a success rate of 100 %.

To validate the optimization results, SGE and GE were compared using the statistical tools previously described. The outcomes presented in the column Statistical Validation of Table 2 reveal that SGE provides statistical significant improvements over the standard GE. We present the p-values obtained, to clarify the magnitude of the differences. The highest p-value is the one for the harmonic experiment, and it still is far from the $\alpha = 0.05$ that was selected as level of

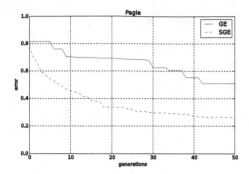

Fig. 5. Mean Best Fitness plots for the Pagie polynomial

Table 2. Optimization results: Mean Best Fitness and standard deviation over 30 runs

Problem	GE	SGE	Statistical validation	
			p-value	Effect Size
Harmonic curve regression	0.20 ($\pm$ 0.11)	0.13($\pm$ 0.05)	$6.09 * 10^{-3}$	++
Pagie polynomial	0.50 ($\pm$ 0.26)	0.29 ($\pm$ 0.09)	$2.20 * 10^{-6}$	+++
Santa Fe Ant trail	21.40 ($\pm$ 12.40)	0.00 ($\pm$ 0.00)	$9.45 * 10^{-11}$	+++

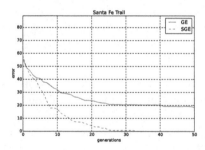

Fig. 6. Mean Best Fitness plots for the Santa Fe Ant trail

significance. We also computed the effect sizes, to assess how large the differences were. The only problem where the effect size is medium ($0.3 <= r < 0.5$) is the harmonic number. In all other problems the effect size is large. These results suggest that SGE is a valid alternative to GE.

5   Conclusion

In this paper we proposed Structured Grammatical Evolution (SGE), a new genotypic representation for GE that explicitly considers the features of the grammar being used. The definition of the genotype requires two pre-processing steps: first, recursive productions are rewritten in a non-recursive format, which requires the addition of several new non-terminals; then, an upper bound for the maximum number of non-terminals expansion is computed. After pre-processing is over, the structured genotype is defined. Each gene links to a specific non-terminal and it encodes a list of integers that help to determine the derivation options during mapping. SGE effectiveness was tested on a set of benchmarks problems and results were encouraging, as it was able to outperform the standard GE representation in all selected problems. Moreover, it proved to be efficient, as it needed a lower number of evaluations to discover good quality solutions.

Standard GE has been criticized due to the low locality and extremely high redundancy [9]. One of the goals of the representation proposed in this paper is to enhance GE with a valuable tool to handle these two limitations. We are currently performing a comprehensive set of empirical tests focused on locality. Preliminary results are promising, as they confirm that SGE has higher locality than standard GE [5]. In the near future we will extend the analysis, in order to gain a deeper insight on how SGE impacts locality and redundancy.

Acknowledgments. This work was partially supported by Fundação para a Ciência e Tecnologia (FCT), Portugal, under the grant SFRH/BD/79649/2011.

References

1. Fagan, D., O'Neill, M., Galván-López, E., Brabazon, A., McGarraghy, S.: An analysis of genotype-phenotype maps in grammatical evolution. In: Esparcia-Alcázar, A.I., Ekárt, A., Silva, S., Dignum, S., Uyar, A.Ş. (eds.) EuroGP 2010. LNCS, vol. 6021, pp. 62–73. Springer, Heidelberg (2010)
2. Field, A.P.: How to Design and Report Experiments. SAGE, Chicago (2003)
3. Keijzer, M., O'Neill, M., Ryan, C., Cattolico, M.: Grammatical evolution rules: the mod and the bucket rule. In: Foster, J.A., Lutton, E., Miller, J., Ryan, C., Tettamanzi, A.G.B. (eds.) EuroGP 2002. LNCS, vol. 2278, pp. 123–130. Springer, Heidelberg (2002)
4. Koza, J.R.: Genetic Programming: On the Programming of Computers by Means of Natural Selection. MIT Press, Cambridge, MA, USA (1992)
5. Lourenço, N., Pereira, F.B., Costa, E.: An inquiry into the properties of structured grammatical evolution. Technical report, ECOS-CISUC (2014)
6. O'Neill, M., Ryan, C.: Grammatical Evolution: Evolutionary Automatic Programming in an Arbitrary Language. Kluwer Academic Publishers, Norwell (2003)
7. O'Neill, M., Brabazon, A., Nicolau, M., Garraghy, S.M., Keenan, P.: ΠGrammatical Evolution. In: Deb, K., Tari, Z. (eds.) GECCO 2004. LNCS, vol. 3103, pp. 617–629. Springer, Heidelberg (2004)
8. O'Neill, M., Hemberg, E., Gilligan, C., Bartley, E., McDermott, J., Brabazon, A.: GEVA - grammatical evolution in Java. Technical report (2008)
9. Rothlauf, F., Oetzel, M.: On the locality of grammatical evolution. In: Collet, P., Tomassini, M., Ebner, M., Gustafson, S., Ekárt, A. (eds.) EuroGP 2006. LNCS, vol. 3905, pp. 320–330. Springer, Heidelberg (2006)
10. Ryan, C., Azad, R.M.A., Sheahan, A., O'Neill, M.: No coercion and no prohibition, a position independent encoding scheme for evolutionary algorithms - the chorus system. In: Foster, J.A., Lutton, E., Miller, J., Ryan, C., Tettamanzi, A.G.B. (eds.) EuroGP 2002. LNCS, vol. 2278, pp. 131–141. Springer, Heidelberg (2002)

11. Urbano, P., Georgiou, L.: Improving grammatical evolution in santa fe trail using novelty search. Adv. Artif. Life ECAL **12**, 917–924 (2013)
12. White, D.R., McDermott, J., Castelli, M., Manzoni, L., Goldman, B.W., Kronberger, G., Jaśkowski, W., O'Reilly, U.M., Luke, S.: Better gp benchmarks: community survey results and proposals. Genet. Programm. Evolvable Mach. **14**(1), 3–29 (2013)

Greedy Semantic Local Search for Small Solutions

Robyn Ffrancon and Marc Schoenauer[(✉)]

TAO Project-team, INRIA Saclay - Île-de-France, Université Paris-Sud,
91128 Orsay Cedex, France
rffrancon@gmail.com, Marc.Schoenauer@inria.fr

Abstract. Semantic Backpropagation (SB) was introduced in GP so as to take into account the semantics of a GP tree at all intermediate states of the program execution, i.e., at each node of the tree. The idea is to compute the optimal "should-be" values each subtree should return, whilst assuming that the rest of the tree is unchanged, and to choose a subtree that matches as well as possible these target values. A single tree is evolved by iteratively selecting and replacing a single node with the best subtree from a static library. Replacements are made with the primary aim of reducing the local error, and a secondary aim of reducing the tree size. Previous results for standard Boolean GP benchmarks that have been obtained by the authors with another variant of SB are improved in term of tree size. SB is then applied for the first time to categorical GP benchmarks, and outperforms all known results to date for three variable finite algebras.

1 Introduction

Local search algorithms are generally the most straightforward optimization methods that can be designed on any search space that has some neighborhood structure. Given a starting point (usually initialized using some randomized procedure), the search proceeds by selecting the next point, from the neighborhood of the current point, which improves the value of the objective function, with several possible variants (e.g., first improvement, best improvement, etc.). When the selection is deterministic, the resulting *Hill Climbing* algorithms generally perform poorly, and rapidly become intractable on large search spaces. Stochasticity must be added, either to escape local minima (e.g. through restart procedures from different random initializations, or by sometimes allowing the selection of points with worse objective value than the current point), or to tackle very large search spaces (e.g., by considering only a small part of the neighborhood of the current point). The resulting algorithms, so-called *Stochastic Local Search* algorithms (SLS) [2], are today the state-of-the-art methods in many domains of optimization.

The concept of a neighborhood can be equivalently considered from the point of view of some *move* operators in the search space: the neighborhood of a point is the set of points which can be reached by application of that *move*

ⓒ Springer International Publishing Switzerland 2016
S. Bonnevay et al. (Eds.): EA 2015, LNCS 9554, pp. 149–162, 2016.
DOI: 10.1007/978-3-319-31471-6_12

operator. This perspective encourages the use of stochasticity in a more flexible way by randomizing the *move* operator, thus dimming the boundary between local and global search. It also allows the programmer to introduce domain specific knowledge in the operator design.

All $(1^+_+\lambda)$-EAs can be viewed as Local Search Algorithms, as the mutation operator acts exactly like the *move* operator mentioned above. The benefit of EAs in general is the concept of population, which permits the transfer of more information from one iteration to the next. However in most domains, due to their simplicity, SLS algorithms have been introduced and used long before more sophisticated metaheuristics like Evolutionary Algorithms (EAs). But this is not the case in the domain of Program Synthesis[1] where Genetic Programming (GP) was the first algorithm related to Stochastic Search which took off and gave meaningful results [3]. The main reason for that is probably the fact that performing random moves on a tree structure rarely results in improving the objective value (aka fitness, in EA/GP terminology).

Things have begun to change with the introduction of domain-specific approaches to GP, under the generic name of *Semantic GP*. For a given set of problem variable values, the *semantics* of a subtree within a given tree is defined as the vector of values computed by this subtree for each set of input values (each fitness case). In Semantic GP, as the name implies, the semantics of all subtrees considered as well as the semantics of the context in which a sub-tree is inserted (i.e., the semantics of its siblings), as first proposed and described in detail in [6] (see also [10] for a recent survey). Several variation operators have been proposed for use within the framework of Evolutionary Computation (EC) which take semantics into account when choosing and modifying subtrees. In particular, *Semantic Backpropagation* (SB) [5,7,11] were the first works to take into account not only the semantic of a subtree to measure its potential useful-ness, but also the semantics of the target node where it might be planted. The idea of SB was pushed further in [1], a paper published by the authors, where the first (to the best of our knowledge) Local Search algorithm, called Iterated Local Tree Improvement (ILTI), was proposed and experimented with on stan-dard Boolean benchmark problems for GP. Its efficiency favorably compared to previous works (including Behavioural Programming GP [4], another successful approach to learn the usefulness of subtrees from their semantics using Machine Learning).

The present work[2] builds on [1] in several ways. Firstly, Semantic Backpro-gation is extended from Boolean to categorical problems. Second, and maybe more importantly, the algorithm itself is deeply modified and becomes Iterated Greedy Local Tree Improvements (IGLTI): On one hand, the library from which replacement subtrees are selected usually contain all possible depth-k subtrees ($k = 2$ or $k = 3$). On the other hand, during each step of the algorithm, a strong emphasis is put on trying to minimize the total size of the resulting tree.

---

[1] See also [8] for a survey on recent program synthesis techniques from formal methods and inductive logic programming, to GP.

[2] That will also be presented at the Semantic Workshop (SMGP) at GECCO 2015.

Indeed, a modern interpretation of the Occam's razor principle states that small solutions should always be preferred to larger ones – the more so in Machine Learning in general, where large solutions tend to learn "by heart" the training set, with poor generalization properties. And this is even more true when trying to find an exact solution to a (Boolean or categorical) problem with GP. For instance in the categorical domain of finite algebras (proposed in [9]), there exist proven exact methods for generating the target terms. However these methods generate solutions with millions of terms that are of little use to mathematicians.

The paper is organized as follows: Sect. 2 recalls the basic idea of Semantic Backpropagation, illustrated in the categorical case here. Section 3 then describes in detail the new algorithm IGLTI. Section 4 introduces the benchmark problems, again concentrating on the categorical ones, and Sect. 5 presents the experimental results of IGLTI, comparing them with those of the literature as well as those obtained by ILTI [1]. Finally Sect. 6 concludes the paper, discussing the results and sketching further directions of research.

2   Semantic Backpropagation

2.1   Hypotheses and Notations

The context is that of supervised learning: The problem at hand comprises n fitness cases, were each case i is a pair (x_i, f_i), x_i being a vector of values for the problem variables, and f_i the corresponding desired tree output. For a given a loss function ℓ, the goal is to find the program (*tree*) that minimizes the global error

$$Err(tree) = \sum_{i=1}^{i=n} \ell(tree(x_i), f_i) \tag{1}$$

where $tree(x_i)$ is the output produced by the tree when fed with values x_i.

In the Boolean framework, for instance, each input x_i is a vector of Boolean variables, and each output f_i is a Boolean value. A trivial loss function is the Hamming distance between Boolean values, and the global error of a tree is the number of errors of that tree.

2.2   Rationale

The powerful idea underlying Semantic Backpropagation is that, for a given tree, it is very often possible to calculate the optimal outputs of each node such that the final tree outputs are optimized. Each node (and rooted subtree) is analyzed under the assumption that the functionality of all the other tree nodes are optimal. In effect, for each node, the following question should be asked: What are the optimal outputs for this node (and rooted subtree) such that its combined use with the other tree nodes produce the optimal final tree outputs? Note that for any given node, its optimal outputs do not depend on its semantics (actual outputs). Instead, they depend on the final tree target outputs, and the actual output values (semantics) of the other nodes within the tree.

In utilizing the results of this analysis, it is possible to produce local fitness values for each node by comparing their actual outputs with their optimal outputs.

Similarly, a fitness value can be calculated for any external subtree by comparing its actual outputs to the optimal outputs of the node which it might replace. If this fitness value indicates that the external subtree would perform better than the current one, then the replacement operation should improve the tree as a whole.

In the following, we will be dealing with a *master tree* T and a *subtree library* $\mathcal{L}$. We will now describe how a subtree (node location) s is chosen in T **together with** a subtree s^* in $\mathcal{L}$ to try to improve the global fitness of T (aggregation of the error measures on all fitness cases) when replacing, in T, s with s^*.

2.3   Tree Analysis

For each node in T, the GLTI algorithm maintains an *output vector* and an *optimal vector*. The ith component of the output vector is the actual output of the node when the tree is executed on the ith fitness case; the ith component of the optimal vector is the value that the node should take so that its propagation upward would lead T to produce the correct answer for this fitness case, all other nodes being unchanged.

The idea of storing the output values is one major component of BPGP [4], which is used in the form of a trace table. In their definition, the last column of the table contained target output values of the full tree – a feature which is not needed here as they are stored in the optimal vector of the root node.

Let us now detail how these vectors are computed. The output vector is simply filled during the execution of T on the fitness cases. The computation of the optimal vectors is done in a top-down manner. The optimal values for the top node (the root node of T) are the target values of the problem. Consider now a simple tree with top node A and child nodes B and C. For a given fitness case, denote by a, b and c their respective returned values, and by $\hat{a}$, $\hat{b}$ and $\hat{c}$ their optimal values (or set of optimal values, see below)[3]. Assuming now that we know $\hat{a}$, we want to compute $\hat{b}$ and $\hat{c}$ (top-down computation of optimal values).

If node A represents operator F, then, by definition

$$a = F(b, c) \tag{2}$$

and we want $\hat{b}$ and $\hat{c}$ to satisfy

$$\hat{a} = F(\hat{b}, c) \text{ and } \hat{a} = F(b, \hat{c}) \tag{3}$$

i.e., to find the values such that A will take a value $\hat{a}$, assuming the actual value of the other child node is correct. This leads to

$$\hat{b} = F_b^{-1}(\hat{a}, c) \text{ and } \hat{c} = F_c^{-1}(\hat{a}, b) \tag{4}$$

---

[3] The same notation will be implicit in the rest of the paper, whatever the nodes A, B and C.

where F_k^{-1} is the pseudo-inverse operator of F which must be used to obtain the optimum $\hat{k}$ of variable k. The definition of the pseudo-inverse operators in the Boolean case is simpler than that in the categorical case. Only the latter will be discussed now – see [1] for the Boolean case.

Firstly, in the Boolean case, all operators are symmetrical - hence F_b^{-1} and F_c^{-1} are identical. However, in the categorical problems considered here, the (unique) operator is not commutative (i.e., the tables in Fig. 1 are not symmetrical), hence F_b^{-1} and F_c^{-1} are different.

Secondly, the pseudo-inverse operator is multivalued: for example, from inspecting the finite algebra $A4$ (Fig. 1-left), it is clear to see that if $\hat{a} = 1$ and $b = 0$ then $\hat{c}$ must equal 0 or 2. In which case we write $\hat{c} = (0, 2)$. That is to say, if $c \in \hat{c}$ and $b = 0$ then $a = 1$. For this example, the pseudo-inverse operator is written as $F_c^{-1}(1, 0) = (0, 2)$. On the other hand, from Fig. 1-right, it comes that $F_b^{-1}(1, 0) = 0$.

Now, consider a second example where the inverse operator is ill-defined. Suppose $\hat{a} = 1$, $b = 1$, and we wish to obtain the value of $\hat{c} = F_c^{-1}(1, 1)$. From inspecting row $b = 1$ of $A4$ we can see that it is impossible to obtain $\hat{a} = 1$ regardless of the value of c. Further inspection reveals that $\hat{a} = 1$ when $b = 0$ and $c = (0, 2)$, or when $b = 2$ and $c = 1$.

Therefore, in order to chose $\hat{c}$ for $\hat{a} = 1$ and $b = 1$, we must assume that $b = 0$ or that $b = 2$. If we assume that $b = 2$ we then have $\hat{c} = 1$. Similarly, if we assume that $b = 0$ we will have $\hat{c} = (0, 2)$. The latter assumption is preferable because we assume that it is less likely for c to satisfy $\hat{c} = 1$ than $\hat{c} = (0, 2)$. In the latter case, c must be one of two different values (namely $c = 0$ or $c = 2$) where as in the former case there is only one value which satisfies $\hat{c}$ (namely $c = 1$). We therefore choose $F_c^{-1}(1, 1) = (0, 2)$. However, as a result, we must also have $F_b^{-1}(1, 0) = 0$ and $F_b^{-1}(1, 2) = 0$.

Of course, for the sake of propagation, the pseudo-inverse operator should also be defined when $\hat{a}$ is a tuple of values. For example, consider the case when $\hat{a} = (1, 2)$, $c = 0$, and $\hat{b}$ is unknown. Inspecting column $c = 0$ in $A4$ will reveal that the only a value that will satisfy $\hat{a}$ (namely $a = 1$ satisfies $\hat{a} = (1, 2)$) is found at row $b = 1$. Therefore, in this case $\hat{b} = F_b^{-1}((1, 2), 0) = 1$.

Using the methodologies outlined by these examples it is possible to derive pseudo-inverse function tables for all finite algebras considered in this paper. As an example, Fig. 2 gives the complete pseudo-inverse table for finite algebra $A4$.

Having defined the pseudo-inverse operators, we can compute, for each fitness case, the optimal vector for all nodes of T, starting from the root node and computing, for each node in turn, the optimal values for its two children as described above, until reaching the terminals.

2.4   Local Error

The local error of each node in T is defined as the discrepancy between its output vector and its optimal vector. The loss function ℓ that defines the global error from the different fitness cases (see Eq. 1) can be reused, provided that it is

A4	c 0	1	2
0	1	0	1
b 1	0	2	0
2	0	1	0

B1	c 0	1	2	3
0	1	3	1	0
b 1	3	2	0	1
2	0	1	3	1
3	1	0	2	0

Fig. 1. Function tables for the primary algebra operators $A4$ and $B1$.

$\hat{a}$	b	$\hat{c}$	$\hat{a}$	c	$\hat{b}$
0	0	1	0	0	(1,2)
	1	(0,2)		1	0
	2	(0,2)		2	(1,2)
1	0	(0,2)	1	0	0
	1	(0,2)		1	2
	2	1		2	0
2	0	1	2	0	1
	1	1		1	1
	2	1		2	1
(0,1)	0	(0,1,2)	(0,1)	0	(0,1,2)
	1	(0,2)		1	(0,2)
	2	(0,1,2)		2	(0,1,2)
(0,2)	0	1	(0,2)	0	(1,2)
	1	(0,1,2)		1	(0,1)
	2	(0,2)		2	(1,2)
(1,2)	0	(0,2)	(1,2)	0	0
	1	1		1	(1,2)
	2	1		2	0
(0,1,2)	0	(0,1,2)	(0,1,2)	0	(0,1,2)
	1	(0,1,2)		1	(0,1,2)
	2	(0,1,2)		2	(0,1,2)

Fig. 2. Pseudo-inverse operator function tables for the $A4$ categorical benchmark.

extended to handle sets of values. In the categorical context, the distance between different values should be independent of the values themselves ($d(a, a) = 0$ and $d(a, b) = 1$ if $a \neq b$). This leads to the following extension of this standard Hamming-like distance (for both the Boolean and categorical contexts): Denoting the output and optimal values for node A on fitness case i as a_i and $\hat{a}_i$ respectively, the local error $Err(A)$ of node A is defined as

$$Err(A) = \sum_i \ell(a_i, \hat{a}_i) \tag{5}$$

were

$$\ell(a_i, \hat{a}_i) = \begin{cases} 0, & \text{if } a_i \in \hat{a}_i \\ 1, & \text{otherwise.} \end{cases} \tag{6}$$

2.5 Subtree Library

Given a node A in T that is candidate for replacement (see next Sect. 3.1 for possible strategies for choosing it), we need to select a subtree in the library $\mathcal{L}$

that would likely improve the global fitness of T if it were to replace A. Because the effect of replacement on the global fitness is, in general, beyond the scope of this investigation, we have chosen to use the local error of A as a proxy. Therefore, we need to compute the *substitution error* $Err(B, A)$ of any node B in the library, i.e. the local error of node B if it were inserted in lieu of node A. Such error can obviously be written as

$$Err(B, A) = \sum_i \ell(b_i, \hat{a}_i) \tag{7}$$

Then, for a given node A in T, we can find $best(A)$, the set of subtrees in $\mathcal{L}$ with minimal substitution error,

$$best(A) = \{B \in \mathcal{L}; Err(B, A) = min_{C \in \mathcal{L}}(Err(C, A))\} \tag{8}$$

and then define the *Expected Local Improvement* $I(A)$ as

$$I(A) = Err(A) - Err(B, A) \text{ for some } B \in best(A) \tag{9}$$

If $I(A)$ is positive, then replacing A with any node in $best(A)$ will improve the local fitness of A. Note however that this does not imply that the global fitness of T will improve. Indeed, even though the local error will decrease, the erroneous fitness cases may differ, which could adversely affect the whole tree. On the other hand, if $I(A)$ is negative, no subtree in $\mathcal{L}$ can improve the global fitness when inserted in lieu of A.

Two different IGLTI schemes were tested on the categorical benchmarks which we will refer to as: IGLTI depth 2 and IGLTI depth 3. In the IGLTI depth 2 scheme the library consisted of all semantically unique trees from depth 0 to depth 2 inclusive. Similarly, in the IGLTI depth 3 scheme all semantically unique trees from depth 0 to depth 3 were included. Only the IGLTI depth 3 scheme was tested on the Boolean benchmarks. In this case, the library size was limited to a maximum of 40000 trees.

The library for the ILTI algorithm was constructed from all possible semantically unique subtrees of 2500 randomly generated full trees of depth 2. In this case the library had a strict upper size limit of 450 trees and the library generating procedure immediately finished when this limit was met. Note that for the categorical benchmarks, the size of the library was always below 450 trees. For the Boolean benchmarks on the other hand, the library size was always 450 trees.

In the process of generating the library (whatever design procedure is used), if two candidate subtrees have exactly the same outputs, only the tree with fewer nodes is kept. In this way, the most concise generating tree is stored for each output vector. The library $\mathcal{L}$ is ordered by tree size, from smallest to largest, hence so is $best(A)$. Table 1 gives library sizes for each categorical benchmarks.

Algorithm 1. Procedure GLTI(Tree T, library $\mathcal{L}$)

Require: $Err(A)$ (Eq. 5), $Err(B, A)$ (Eq. 7), $A \in T$, $B \in \mathcal{L}$

 1   $\mathcal{T} \leftarrow \{A \in T;\ \textbf{if}\ Err(A) \neq 0\}$

 2   $bestErr \leftarrow +\infty$
 3   $bestReduce \leftarrow +\infty$
 4   $bestANodes \leftarrow \{\}$

 5   **for** $A \in \mathcal{T}$ **do**   ▷ Loop over nodes which could be improved
 6      $A.minErr \leftarrow +\infty$
 7      $A.minReduce \leftarrow +\infty$
 8      $A.libraryTrees \leftarrow \{\}$
 9      $indexA \leftarrow$ index position of A in tree T

10   **for** $B \in \mathcal{L}$ **do**   ▷ Loop over trees in library
11   **if** $B \in T.bannedBTrees(indexA)$ **then**
12   **continue**

13          $BReduce \leftarrow \text{size}(B) - \text{size}(A)$

14   **if** $Err(B, A) < A.minErr$ **then**
15             $A.minErr \leftarrow Err(B, A)$
16             $A.minReduce \leftarrow BReduce$
17             $A.libraryTrees \leftarrow \{B\}$

18   **if** $Err(B, A) = 0$ **then**
19   **break**   ▷ Stop library search for current A

20   **else if** $Err(B, A) = A.minErr$ **then**
21   **if** $BReduce < A.minReduce$ **then**
22                $A.minReduce \leftarrow BReduce$
23                $A.libraryTrees \leftarrow \{B\}$

24   **else if** $BReduce = A.minReduce$ **then**
25                $A.libraryTrees.\text{append}(B)$

26   **if** $A.minErr < bestErr$ **then**
27          $bestErr \leftarrow A.minErr$
28          $bestReduce \leftarrow A.minReduce$
29          $bestANodes \leftarrow \{A\}$

30   **else if** $A.minErr = bestErr$ **then**
31   **if** $A.minReduce < bestReduce$ **then**
32             $bestReduce \leftarrow A.minReduce$
33             $bestANodes \leftarrow \{A\}$

34   **else if** $A.minReduce = bestReduce$ **then**
35             $bestANodes.\text{append}(A)$

36   $chosenA \leftarrow random(bestANodes)$
37   $chosenB \leftarrow random(chosenA.libraryTrees)$

38   $indexA \leftarrow$ index position of $chosenA$ in T
39   $T.bannedBTrees(indexA).\text{append}(chosenB)$

40   **return** $chosenA$, $chosenB$, T

Table 1. Library sizes for each categorical benchmark.

Benchmark	Library size		
	IGLTI depth 3	IGLTI depth 2	ILTI depth 2
D.A1	16945	138	72
D.A2	19369	144	78
D.A3	18032	145	81
D.A4	14963	133	69
D.A5	20591	145	81
M.A1	12476	134	68
M.A2	16244	144	78
M.A3	10387	145	81
M.A4	11424	130	66
M.A5	19766	145	81
M.B	21549	-	81

3 Tree Improvement Procedures

3.1 Greedy Local Tree Improvement

Everything is now in place to describe the full GLTI algorithm, its pseudo-code can be found in Algorithm 1. The algorithm starts (line 1) by storing all nodes $A \in T$ where $Err(A) \neq 0$ in the set $\mathcal{T}$. Then, the nodes in $\mathcal{T}$ are each examined individually (line 5).

The library $\mathcal{L}$ is inspected (lines 14–25) for each node $A \in \mathcal{T}$ with the aim of recording each associated library tree B which firstly minimises $Err(B, A)$ and secondly minimises $BReduce = \text{size}(B) - \text{size}(A)$. In the worst case, for each node A, every tree B within the library $\mathcal{L}$ is inspected. However, the worst case is avoided, and the inspection of the library is aborted, if a tree $B \in \mathcal{L}$ is found which satisfies $Err(B, A) = 0$.

The master tree T can effectively be seen as an array where each element corresponds to a node in the tree. When a library tree B replaces a node and its corresponding rooted subtree in T a record is kept of the index position at which B was inserted. For a node A in the master tree, at line 11 the algorithm ensures that the library trees which have previously been inserted at the T array index position of node A are not considered for insertion again at that index position. This ensures that the algorithm does not become stuck in repeatedly inserting the same B trees to the same array index positions of the master tree T.

After inspecting the library for a given node A, the values $A.minErr$ and $A.minReduce$ are used to determine the set of the very best $A \in \mathcal{T}$ nodes, $bestANodes \subseteq \mathcal{T}$ (lines 26–35).

Next, the algorithm (line 36) randomly chooses a node $chosenA \in bestANodes$ and randomly chooses an associated tree from its best library tree set $chosenB \in chosenA.libraryTrees$.

Finally, the algorithm records the chosen library tree *chosenB* as having been inserted at the array index position of *chosenA* in T.

Complexity. Suppose that the library $\mathcal{L}$ is of size o. The computation of the output vectors of all trees in $\mathcal{L}$ is done once and for all. Hence the overhead of one iteration of GLTI is dominated, in the worst case, by the comparisons of the optimal vectors of all m nodes in T with the output vectors of all trees in $\mathcal{L}$, with complexity $n \times m \times o$.

3.2 Iterated GLTI

In the previous section, we have defined the GLTI procedure that, given a master tree T and a library of subtrees $\mathcal{L}$, selects a node *chosenA* in T and a subtree *chosenB* in $\mathcal{L}$ to insert in lieu of node *chosenA* so as to minimize some local error over a sequence of fitness cases as primary criterion, and the full tree size as secondary criterion. In this section we will turn GLTI into a full Stochastic Search Algorithm.

As discussed in [1], or as done in [7], GLTI could be used within some GP algorithm to improve it with some local search, "à la" memetic. However, following the road paved in [1], we are mainly interested here in experimenting with GLTI a full search algorithm that repeatedly applies GLTI to the same tree. Note that the same tree and the same library will be used over and over, so the meaning of "iterated" here does not involve random restarts. On the other hand, the only pressure toward improving the global fitness will be put on the local fitness defined by Eq. 9. In particular, there are cases where none of the library trees can improve the local error: the smallest decrease is nevertheless chosen, hopefully helping to escape some local optimum.

The parameters of IGLTI are the choice of the initial tree, the method (and its parameters) used to create the library, and the size of the library. The end of the paper is devoted to some experimental validation of IGLTI and the study of the sensitivity of the results w.r.t. its most important parameter, the depth of the library trees.

3.3 Modified ILTI

The ILTI scheme (first introduced in [1]) was modified for use in this paper. In the IGLTI algorithm, a record is kept of which library trees were inserted at each array index positions of the master tree. This feature ensured that the same library tree was not inserted at the same array index positions of the master tree more than once. This feature was also implemented in the modified ILTI scheme. Note that for the rest of this paper the modified ILTI scheme will simply be referred to as the ILTI scheme.

4   Experimental Conditions

The benchmark problems used for these experiments are classical Boolean problems and some of the finite algebra categorical problems which have been proposed in [9] and recently studied in [4,7]. For the sake of completeness, we reiterate their definitions as stated in [4].

"The solution to the *v-bit Comparator problem Cmp-v* must return *true* if the $\frac{v}{2}$ least significant input bits encode a number that is smaller than the number represented by the $\frac{v}{2}$ most significant bits. For the *Majority problem Maj-v*, *true* should be returned if more than half of the input variables are true. For the *Multiplexer problem Mul-v*, the state of the addressed input should be returned (6-bit multiplexer uses two inputs to address the remaining four inputs, 11-bit multiplexer uses three inputs to address the remaining eight inputs). In the *Parity problem Par-v*, *true* should be returned only for an odd number of true inputs.

The categorical problems deal with evolving algebraic terms and dwell in the ternary (or quaternary) domain: the admissible values of program inputs and outputs are $\{0, 1, 2\}$ (or $\{0, 1, 2, 3\}$. The peculiarity of these problems consists of using only one binary instruction in the programming language, which defines the underlying algebra. For instance, for the A4 and B1 algebras, the semantics of that instruction are given in Fig. 1.

For each of the five algebras considered here, we consider two tasks. In the discriminator term tasks, the goal is to synthesize an expression (using only the one given instruction) that accepts three inputs x, y, z and returns x if $x \neq y$ and z if $x = y$. In ternary domain, this gives rise to $3^3 = 27$ fitness cases.

The second task defined for each of algebras consists in evolving a so-called *Mal'cev* term, i.e., a ternary term that satisfies $m(x, x, y) = m(y, x, x) = y$. Hence there are only 15 fitness cases for ternary algebras, as the desired value for $m(x, y, z)$, where x, y, and z are all distinct, is not determined."

In the ILTI algorithm a master tree is initialised as a random full tree of depth 2. For the IGLTI algorithm, the initial master tree is chosen as the best performing subtree from the subtree library. If there are multiple library trees with the same performance, the smallest tree is chosen.

Hard run time limits of 5000 seconds were set for each experiment. A run was considered a failure if a solution was not found within this time.

All results were obtained using an 64bits Intel(R) Xeon(R) CPU X5650 @ 2.67GHz. All of the code was written in Python[4].

5   Experimental Results

Figure 3 shows standard box-plots for solution tree sizes obtained while testing the ILTI and IGLTI depth 3 algorithms on the 6 bit and Cmp08 Boolean benchmarks. It shows how the IGLTI algorithm finds solution trees which are smaller (number of nodes) than those found by the ILTI algorithm. Four failed runs are reported in this figure which occurred when the IGLTI depth 3 algorithm was tested on the Cmp08 benchmark.

The figure also shows how the spread of solution sizes are generally narrower for IGLTI depth 3 than for ILTI. The only exception to this generality is the

---

[4] The entire code base is freely available at robynffrancon.com/GLTI.html.

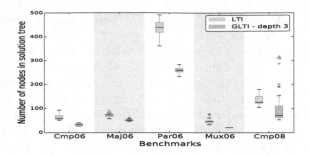

Fig. 3. Standard box-plots for the program solution tree sizes (number of nodes) for the ILTI algorithm and IGLTI depth 3 algorithm tested on the Boolean benchmarks. Each algorithm performed 20 runs for each benchmark. Perfect solutions were found from each run except for the Cmp08 benchmark where the IGLTI algorithm failed 4 times (as indicated by the red number four) (Color figure online).

results of the Cmp08 benchmark. Additional supporting data for this figure is given in Table 2. From inspecting the figure and table together, it is clear that the 20 solution trees obtained from testing IGLTI depth 3 on the Mux06 benchmark were all of the same size.

Figure 4 shows standard box-plots for the number of operators used in the categorical benchmark solutions which were found using the ILTI, IGLTI depth 3, and IGLTI depth 2 schemes. Supporting data for this figure can also be seen in Table 2. However, note that this table measures tree sizes by the number of nodes and not by the number of operators.

The figure shows how the IGLTI depth 3 scheme found the smallest solutions on the $D.A2$, $D.A4$, $D.A5$, $M.A1$, and $M.A2$ benchmarks. For all other three variable categorical benchmarks, the IGLTI depth 2 scheme found the smallest solutions. In all cases, the spread of solution sizes (number of operators) were smallest for IGLTI depth 3 and largest for the ILTI scheme. Reminiscent of the Mux06 benchmark results, the IGLTI depth 3 scheme found twenty solutions which were all of the same size when tested on the $M.A3$ benchmark.

Table 2 gives the algorithm runtimes for each benchmark. The ILTI algorithm is the best performing algorithm for this measure. However, note that the IGLTI depth 2 scheme showed similar average runtimes (but larger spreads) for the three variable $Mal'cev$ term benchmarks.

Nine runs of the ILTI algorithm failed to find a solution within the 5000 second time limit when testing on the $M.B$ benchmark. An average of 387.2 ± 283.0 operators were used per correct solution. The IGLTI depth 3 scheme failed to find a solution once when testing on the $M.B$ benchmark. An average of 88.4 ± 21.4 operators were used by the correct solutions found in this case.

6   Discussion and Further Work

The results presented in this paper show that SB can be successfully used to solve standard categorical benchmarks when the pseudo-inverse functions are carefully

Table 2. Run time (seconds) and program size (number of nodes) results for the ILTI algorithm and IGLTI algorithm (library tree maximum depths 2 and 3) tested on the 6bits Boolean benchmarks and the categorical benchmarks. 20 runs were conducted for each benchmark. The best average results from each row are in bold. The BP4A column is the results of the best performing algorithm of [4]. All runs found perfect solutions with the exception of those indicated by *.

	Run time [seconds]						
	IGLTI - library depth 3		IGLTI - library depth 2		modified ILTI		BP4A
	mean	min	mean	min	mean	min	mean
D.A1	298.1 ± 15.2	272.3	9.8 ± 6.8	2.3	**2.6 ± 1.4**	1.1	136*
D.A2	315.5 ± 18.9	289.6	4.7 ± 2.2	2.0	**1.3 ± 0.3**	0.8	95*
D.A3	302.2 ± 23.0	276.7	4.0 ± 3.0	1.0	**1.2 ± 0.4**	0.7	36*
D.A4	308.0 ± 24.4	268.9	53.5 ± 69.6	4.0	**5.3 ± 3.3**	2.7	180*
D.A5	349.2 ± 39.7	282.6	23.8 ± 9.0	11.9	**3.1 ± 1.6**	0.9	96*
M.A1	191.3 ± 15.7	162.5	1.1 ± 0.6	0.4	**1.0 ± 0.3**	0.5	41*
M.A2	241.2 ± 8.6	230.8	1.0 ± 0.4	0.5	**0.8 ± 0.2**	0.4	21*
M.A3	161.7 ± 7.4	148.0	**0.8 ± 0.3**	0.4	0.9 ± 0.2	0.5	27*
M.A4	171.1 ± 5.7	160.7	3.2 ± 1.3	1.3	**1.0 ± 0.3**	0.5	9
M.A5	298.1 ± 20.8	263.9	1.7 ± 1.1	0.4	**0.9 ± 0.2**	0.5	14
M.B	2772.9* ± 1943	432*	-	-	**843.6* ± 876.2**	4*	-
Cmp06	111.3 ± 23.3	61.4	-	-	**4.1 ± 0.6**	2.9	15
Maj06	95.7 ± 13.0	70.7	-	-	**4.1 ± 0.5**	2.9	36
Par06	258.2 ± 53.2	164.7	-	-	**13.2 ± 2.5**	7.9	233
Mux06	66.1 ± 6.4	48.8	-	-	**4.1 ± 0.8**	2.6	10
	Program size [nodes]						
D.A1	95.3 ± 4.4	91	**80.7 ± 14.1**	55	260.5 ± 122.0	137	134*
D.A2	**65.7 ± 15.9**	41	92.0 ± 18.7	43	144.5 ± 48.1	81	202*
D.A3	65.1 ± 4.4	61	**54.7 ± 6.6**	45	146.1 ± 46.4	79	152*
D.A4	**84.9 ± 10.4**	67	92.6 ± 12.4	67	320.9 ± 84.8	187	196*
D.A5	**64.6 ± 10.8**	47	98.0 ± 23.1	57	238.0 ± 100.1	89	168*
M.A1	**37.8 ± 2.4**	37	46.9 ± 7.9	33	104.4 ± 41.9	43	142*
M.A2	44.8 ± 3.2	33	**44.3 ± 7.7**	33	70.8 ± 18.2	45	160*
M.A3	49.0 ± 0.0	49	**34.8 ± 3.2**	31	143.1 ± 51.0	75	104*
M.A4	**47.9 ± 2.9**	41	49.8 ± 10.9	33	119.5 ± 35.6	61	115
M.A5	37.8 ± 1.8	35	**31.7 ± 13.1**	21	77.1 ± 26.2	33	74
M.B	**179.4* ± 42.3**	95*	-	-	1591.4* ± 1078.6	353*	-
Cmp06	**32.9 ± 5.2**	27	-	-	64.1 ± 11.9	51	156
Maj06	**51.2 ± 3.3**	47	-	-	71.4 ± 7.6	57	280
Par06	**260.0 ± 12.1**	233	-	-	436.0 ± 29.3	361	356
Mux06	**21.0 ± 0.0**	21	-	-	46.3 ± 11.8	33	117

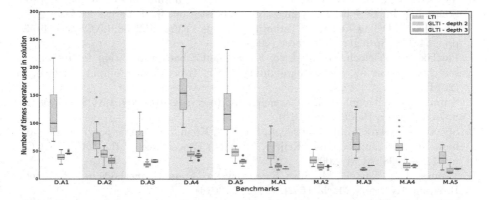

Fig. 4. Standard box-plots for the number of operators in program solutions for the ILTI algorithm and IGLTI algorithm (library tree maximum depths 2 and 3) tested on the categorical benchmarks. Each algorithm performed 20 runs for each benchmark. Perfect solutions were found from each run.

defined. Furthermore, the IGLTI algorithm can be used to find solutions for the three variable categorical benchmarks, which are small enough to be handled by a human mathematician (approximately 45 operators), faster than any other known method.

Interestingly, the results suggest that using a larger library can sometimes lead to worse results (compare the IGLTI depth 2 and IGLTI depth 3 algorithms on the $D.A3$ benchmark for instance). This is likely as a result of the very greedy nature of the IGLTI algorithm. A larger library probably provided immediately better improvements which lead the algorithm away from the very best solutions.

Future work should entail making modification to the IGLTI algorithm so that it is less greedy. In principle, these modifications should be easy to implement by simply adding a greater degree of stochasticity so that slightly worst intermediate results can be accepted. Furthermore, the pseudo-inverse functions should be tested as part of schemes similar to those which feature in [7] with dynamic libraries and a population of potential solutions.

References

1. Ffrancon, R., Schoenauer, M.: Memetic semantic genetic programming. In: Silva, S., Esparcia, A. (eds.) Proceedings of the GECCO. ACM (2015) (To appear)
2. Hoos, H.H., Stützle, T.: Stochastic Local Search. Morgan Kaufmann, San Francisco (2005)
3. Koza, J.R., Programming, G.: On the Programming of Computers by Means of Natural Selection, vol. 1. MIT Press, Cambridge (1992)
4. Krawiec, K., O'Reilly, U.-M., Programming, B.: A broader and more detailed take on semantic GP. In: Arnold, D., et al. (ed.) Proceedings of the GECCO, pp. 935–942. ACM Press (2014)
5. Krawiec, K., Pawlak, T.: Approximating geometric crossover by semantic back-propagation. In: Blum, C., Alba, E. (eds.) Proceedings of the 15th GECCO, pp. 941–948. ACM (2013)
6. McPhee, N.F., Ohs, B., Hutchison, T.: Semantic building blocks in genetic programming. In: O'Neill, M., Vanneschi, L., Gustafson, S., Esparcia Alcázar, A.I., De Falco, I., Della Cioppa, A., Tarantino, E. (eds.) EuroGP 2008. LNCS, vol. 4971, pp. 134–145. Springer, Heidelberg (2008)
7. Pawlak, T., Wieloch, B., Krawiec, K.: Semantic backpropagation for designing search operators in geneticprogramming. IEEE Trans. Evol. Comput. **PP**(99), 1 (2014)
8. Schmid, U., Kitzelmann, E., Plasmeijer, R. (eds.): AAIP 2009. LNCS, vol. 5812, p. 74. Springer, Heidelberg (2010)
9. Spector, L., Clark, D.M., Lindsay, I., Barr, B., Klein, J.: Genetic programming for finite algebras. In: Ryan, C., Keijzer, M. (eds.) Proceedings of the 10th GECCO, pp. 1291–1298. ACM (2008)
10. Vanneschi, L., Castelli, M., Silva, S.: A survey of semantic methods in GP. Genet. Program. Evolvable Mach. **15**(2), 195–214 (2014)
11. Wieloch, B., Krawiec, K., Backwards, R.P.: Instruction inversion for effective search in semantic spaces. In: Blum, C., Alba, E. (eds.) Proceedings of the 15th GECCO, pp. 1013–1020. ACM (2013)

Effects of Cooperation in a Bioinspired Multi-agent Autonomous System for Solving Optimization Problems

Marcus dos Santos[1], Denise Souza[1], Henrique E. Borges[1],
Rogério M. Gomes[1(✉)], and Patrick Siarry[2]

[1] Laboratory of Intelligent Systems, CEFET/MG, Av. Amazonas, 7675,
Belo Horizonte, MG 30510-000, Brazil
{marcusricardo,densouza,henrique,rogerio}@lsi.cefetmg.br
[2] Université Paris Est Créteil, LiSSi, 122 rue Paul Armangot,
94400 Vitry-sur-Seine, France
siarry@u-pec.fr

Abstract. Bimasco - Bioinspired Multi-Agent System for Combinatorial Optimization - consists of an autonomous multi-agent system for solving optimization problems of different classes. This system uses the metaphor of artificial life in which the artificial world represents the search space of a problem, populated by a set of feasible solutions of the problem and grouped into inanimate entities, called regions. Similarly, the world is inhabited by animated entities, agents, each encapsulating one metaheuristic. In this context, this paper introduces an asynchronous and non-deterministic model for the dynamics of interactions among agents and regions, so that it operates as a self-organizing discrete dynamical system. Computational experiments were performed using different classes of combinatorial as well as non-combinatorial optimization problems, including one problem involving a function, which is usually used as benchmark for continuous optimization methods and the knapsack problem. The preliminary results thus obtained show that the dynamics of the implemented model is most effective when there is cooperation between the agents, due to the learning process that occurs from the actions and interactions among them.

Keywords: Bioinspired multi-agent system · Generalized metaheuristics · Optimization

1 Introduction

Due to the growing complexity of optimization problems, the issue of metaheuristics/heuristics hybridization has attracted much attention [2,4,15,18,20,21,24]. These algorithms consist in combining different strategies to better explore the search space and encompass constructive methods, local search strategies, local optima escaping strategies and population-based search. A major difficulty of using these combinations of strategies is that they must be carefully designed,

© Springer International Publishing Switzerland 2016
S. Bonnevay et al. (Eds.): EA 2015, LNCS 9554, pp. 163–176, 2016.
DOI: 10.1007/978-3-319-31471-6_13

which requires a good expertise and knowledge of the problem to be solved [22]. An alternative approach is to dynamically combine different metaheuristics through systems composed of autonomous agents that interact, compete and/or collaborate in the search for better solutions [5,12].

Milano [16] proposed a MultiAGent Metaheuristic Architecture (MAGMA). In this architecture, metaheuristics can be seen as a process resulting from the interactions among different kinds of agents organized in a four-layered conceptual structure. The first three levels are composed of one or more agents, whereas the fourth performs the coordination function of the lower levels. At each level, there are one or more agents: on the first level agents build solutions, on the second level agents improve solutions and on the third level agents provide the high level strategy. The authors showed that a simple hybrid algorithm, called guided restart ILS, can be easily conceived as a combination of existing components in the architecture.

Fernandes and collaborators [7] proposed the Multi-agent Architecture for Metaheuristics (MAM) as a flexible framework to solve different optimization problems without the need of rewriting the algorithms from scratch. In this architecture, each metaheuristic is fully encapsulated in just one agent. The interaction between these agents is indirect, in the sense that an agent does not know any of the other agents, but the coordinator agent. This architecture has been applied to the Vehicle Routing Problem with Time Windows (VRPTW). One of the drawbacks of this architecture is the dependency that any agent has on the coordinator agent, i.e., the agents are not autonomous.

Xie and Liu [23] present the multiagent optimization system (MAOS) for solving the traveling salesman problem (TSP). This system is inspired by nature and supports cooperative search by the self-organization of a group of compact agents situated in an environment with public knowledge sharing. In this architecture, the environment is responsible for providing all knowledge of agents to other agents by means of an interaction center. The environment is also responsible for starting and stopping agents when, for example, a solution is found.

Jin and collaborators [11] modelled a full architecture of agents that cooperate with each other in order to find good solutions for the Capacitated Vehicle Routing Problem. The cooperation among agents occurs asynchronously and through a pool where solutions are placed and/or fetched by agents. Also, the architecture uses parallel programming resources for execution in a distributed way. However, this architecture was proposed and modelled for a specific problem and is not flexible with regard the inclusion of new heuristic techniques.

Aydin [1] developed a multi-agent architecture for solving the Multidimensional Knapsack Problem. The architecture was structured to receive only two metaheuristics: Simulated Annealing (SA) and Particle Swarm Optimization (PSO). Basically, this work consists of a hybrid technique where the SA algorithm provides solutions to PSO algorithm, which in turn is used to generate solutions to the Multidimensional Knapsack Problem.

However, these aforementioned architectures do not seem to be as generic as claimed by the authors since the addition of new metaheuristics and/or the introduction of new problems requires major interventions in the architecture. On the other hand, these frameworks exhibit a high dependency on a controlling agent and therefore can not be seen as an autonomous multi-agent system because any unavailability of that particular agent may interrupt the execution of the system as a whole.

In order to circumvent these obstacles, an alternative approach to the problem of metaheuristics hybridization is proposed. We designed several generalized metaheuristics, local and population-based, to be applied to any optimization problem with any type of representation, encapsulated in an autonomous agent. These algorithms are components of a new multi-agent architecture called Bio-Inspired Multi-Agent System for Combinatorial Optimization (Bimasco). This system uses artificial life as an inspiring metaphor and applies it to build autonomous agents that cooperate to solve a large class of optimization problems. In this case, the artificial world represents the search space of a certain problem, which is populated by a set of feasible solutions of the problem, grouped into inanimate entities, called regions. Similarly, the world is inhabited by animated entities called agents that encapsulate the metaheuristics. In this context, Bimasco models the dynamics of interactions among agents as well as among agents and regions as a self-organized discrete dynamical system. Thus, agents concurrently cooperate with each other, to achieve the best solution to the optimization problem. In this kind of dynamically hybridization metaheuristic system, the best solution emerges from the asynchronous interaction between agents, as will be shown, and is tuned by a set of system parameters. Each set of parameters can be viewed as a different kind of hybridization.

This paper is organized as follows. In Sect. 2 we present the modelling and principal features of the Bimasco architecture. Section 3 shows the computational experiments performed using two different classes of optimization problems from the literature, including one problem involving a function, which is usually used as benchmark for continuous optimization methods and the knapsack problem [8]. Finally, Sect. 4 concludes the paper and presents some relevant extensions of this work.

2 Bimasco Architecture

Bimasco architecture presents a mathematical model of an artificial world. This artificial world is composed of three main structures: Agent, Region and Environment. The solutions for optimization problems are present in these structures. Figure 1 illustrates the entities present in the artificial world Bimasco architecture.

Considering optimization problems, the environment structure represents the whole artificial world and can be described as an abstract space S consisting of feasible solutions of a given problem where different autonomous entities coexist. Therefore, in this environment, or search space, there are a number of inanimate entities called regions (R). A region R is a non-empty subset of the search space assembled according to some criterion.

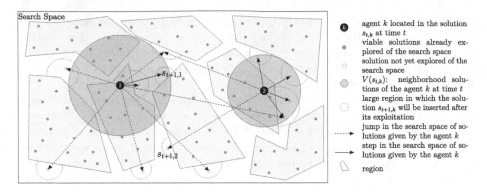

Fig. 1. Representation of the artificial world with some elements: agent, regions and solutions

Given k-regions represented by R^k, we have:

$$\bigcap_{i=1}^{k} R^i = \varnothing \qquad and \qquad \bigcup_{i=1}^{k} R^i = S, \tag{1}$$

i.e., a solution $s \in S$ belongs to a specific region and the union of all the k-regions should necessarily result in the whole search space of the problem. Thus, the aim of the implementation of regions is to build a system in which the solutions could be grouped according to certain common characteristics. This approach contributes to information exchange among entities in the architecture.

Regions are dynamic and can perform the following types of operations: expansion, contraction, fusion and fission. These operations can be triggered by interactions with other entities or by their own internal dynamics. Fission operation occurs, for example, by means of an internal analysis. Thus, the region comprises its own solutions and performs a calculation of the variance of the objective function values of them. If the variance is greater than a preset limit or value, the region will be divided into three other regions, as can be seen in Fig. 2. Otherwise, if two regions identify certain similarity in their solutions, they can be merged into a single region. The union or fusion between two regions is based on an analysis of covariance between the distributions corresponding to the values of the objective function for the set of solutions of each region analysed. If the result of this analysis is lower than a fusion threshold, then a merging operation will be carried out between both regions.

Animated objects in this metaphor of artificial life are autonomous software agents living in this artificial world. These agents remain adapted to their environment through their actions and interactions with other entities. Furthermore, they are provided with motor skills that enable them to get around their environment. At each instant of time t, the agent k is located in a point $s_{t,k}$, i.e., a solution of the search space of the problem. Note that the search space is discrete and the agent moves from one solution to another.

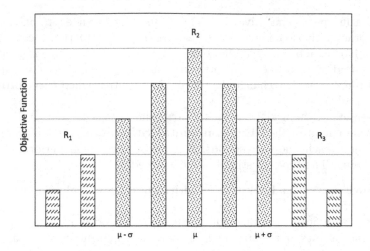

Fig. 2. Region fission strategy based on variance

Figure 1 shows how the agents move in the search space. Basically, agents can perform two types of movements, *step* and *jump*. In step movement, for example, Agent 1 which is located in the solution $s_{t,1}$ in time t, steps toward an neighbour solution, i.e., $s_{t+1,1} \in V(s_{t,1})$. In the jump movement, in turn, Agent 2, which is located in $s_{t,2}$ in time t, takes a jump toward a solution out of its neighbourhood, i.e., $s_{t+1,2} \notin V(s_{t,2})$.

An agent incorporates a heuristic/metaheuristic and the environment is the problem to be solved. The world may have a different set of species of co-evolving creatures. A solution of a different optimization problem basically involves the consideration of a new artificial world. Three different types of agents have been implemented in the architecture:

- Constructor - contains a constructive metaheuristic algorithm for generating initial solutions. For example, GRASP (Greedy Randomized Adaptive Search Procedure) [6].
- Non-Populational - contains a metaheuristic algorithm for generating single solutions. In addition, these agents can receive an initial solution from other agents. For example, ILS (Iterated Local Search) [14] and SA (Simulated Annealing) [13].
- Populational - contains a metaheuristic algorithm for generating a set of solutions. Furthermore, these agents can receive a set of initial solutions from other agents. For example, GA (Genetic Algorithm) [10].

Constructor agents are independent and generate on demand solutions of other agents. On the other hand, the other two types of agents are not constructors and, therefore, they require one or more initial solutions to execute and generate their own solutions. Thus, non-constructors agents request solutions from other agents and/or regions. However, in some cases, agents can request partial solutions from agents that are still in the creation process of their own solutions, thus promoting diversity. Thus, Bimasco is an environment which allows interaction between agents and regions. These interactions occur through stimuli exchange, allowing that the solutions generated by the agents are classified and shared by all of them. Agents communicate among themselves and with regions through stimuli exchange that occurs in an entity called **Environmental Stimuli Pool**.

The stimuli exchanged between the artificial beings (Agent or Region) of the environment are called **Environmental Stimuli**. There are many types of stimuli and each one corresponds to a particular action performed by the artificial being. The types of stimuli are described below:

- **Acknowledge** - stimulus sent to all entities of the artificial world when a new entity is created.
- **RequestSolution** - stimulus sent by the agent whenever it requests solutions from other agents or regions. This stimulus can be sent in broadcast (to all entities) or specifically to a target entity.
- **SendSolution** - stimulus sent in response to a stimulus type **Request-Solution**. **SendSolution** stimulus sends one solution or a set of solutions requested. This type of stimulus is always sent to a specific target, that is, to entities which requested the solution.
- **MergeRequest** - stimulus sent from one region to another requesting a merging.
- **MergeAnswer** - stimulus sent from one region to another in response to a **MergeRequest** stimulus. The target region receives a stimulus of **MergeRequest** type and evaluates the possibility to merge. If possible, the region sends a response stimulus authorizing the completion of the merging.

3   Experimental Results

Computational experiments were performed using different classes of optimization problems, including one problem involving a function, which is usually used as benchmark for continuous optimization methods, and one combinatorial optimization problem, the multidimensional 0-1 knapsack problem [8]. For each of these problems, experiments were performed in two different ways, with and without the possibility of cooperation among agents. Throughout the experimental phase, specific combinations of the following metaheuristic agents were used: GRASP, ILS, SA, VNS and GA.

To run an optimization problem in Bimasco we used two parameter files. The first corresponds to the instance of the problem and should contain the information necessary to define the problem. The other file is related to the architecture configuration parameters, for agents and metaheuristics. The architecture configuration consists of the following parameters: Simulation Name; Experiment Description, Simulated Problem, Fission and Fusion Values of the regions, Instance of the Problem. The agent configuration, in turn, consists of the following parameters: Number of agents, Agent Name, Life Time and Type of Metaheuristic.

Furthermore, metaheuristics to be executed in Bimasco architecture may also require the setup of their solution modifiers. Modifiers are basic implementations of methods that modify or manipulate one or more solutions and that can be shared by different metaheuristics [19]. Some of these parameters were defined in a similar way for all experiments.

The metaheuristics GRASP, ILS, SA and VNS use a **Local Search** solution modifier in common, which can be performed in three different ways: Random Search (RS), First Improvement Local Search and Best Neighbour. GRASP also uses the solution modifier **List of Candidates** of the type Step and Position. ILS uses the **Perturbation** modifier of four different ways: Random, Position, Real-location and Mixed. GA, in turn, uses two modifiers: Mutation and Crossover. The function of the solution modifier mutation in the genetic algorithm is similar to the function of the **Perturbation** modifier of the ILS and therefore it also uses the aforementioned techniques. The metaheuristic VNS uses **Neighbor-hood** modifier which can be performed in two ways: Position and Reallocation. Finally, the metaheuristic SA uses the **Temperature** modifier with the Classical Method and By Iteration.

Metaheuristics implemented in Bimasco have four run-stopping criteria: maximum number of iterations, maximum number of iterations without improvement, runtime and value of the objective function. Each agent can use one criterion or any combination of these criteria. All experiments performed in this work were carried out by using the criterion of maximum number of iterations.

3.1   Experiment 1 - XinSheYang03 Function

The XinSheYang03 function is a benchmark widely used in optimization, particularly when it is desired to compare the performance of different algorithms [9].

Equation (2) defines the XinSheYang03 function while Eq. (3) shows the domain of the variables of the problem and the β and m values. The global minimum of this function, obtained analytically, occurs in $x_1 = 0$ and $x_2 = 0$, for a value of $f(x) = -1$. Figure 3 illustrates graphically this function.

$$f(x) = e^{-\sum_{i=1}^{n}(x_i/\beta)^{2m}} - 2e^{-\sum_{i=1}^{n}(x_i)^2} \prod_{i=1}^{n}\cos^2(x_i) \qquad (2)$$

$$x_i \in [-20, 20], \quad \beta = 15, \quad m = 3 \qquad (3)$$

This experiment aims at evaluating the performance of the architecture for a continuous optimization problem. The architecture was executed with 12 agents with and without cooperation in 30 runs of 15 min each. The agents used were: 4 GRASP, 4 ILS and 4 GA. Table 1 shows the empirically established parametrization of the metaheuristic agents used in this experiment.

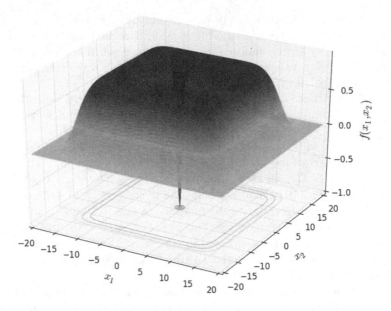

Fig. 3. XinSheYang03 function

Table 1. Metaheuristics Parameters - XinSheYang03 Function

Number of Agents	Metaheuristics	Solution modifers	Parameter values
4	GRASP	Local Search: Best Neighbour List of Candidates: Step	% of Randomness= 0.5 Nº of Iterations = 10
2	ILS	Local Search: Random Perturbation: Random	Perturbation Level = 7 Nº of Iterations = 200
2	ILS	Local Search: Best Neighbour Perturbation: Position	Perturbation Level = 10 Nº of Iterations = 100
2	GA	Crossover: Linear Combination Mutation: Random	Number of Parents = 2 Nº of Iterations = 100 Population Size = 30 Mutation Rate = 0.2 Crossover Rate = 0.7
2	GA	Crossover: Linear Combination Mutation: Position	Number of Parents = 2 No of Iterations = 200 Population Size = 20 Mutation Rate = 0.1 Crossover Rate = 0.8

The results obtained are summarized in Table 2 and show the average values over 30 runs of the architecture with and without cooperation between agents. As can be seen, the average of the values of the objective function obtained with the cooperation in all executions, indicates that the system converges toward a value closer to the optimal compared to the system without the cooperation.

Table 2. Experiment 1: XinSheYang03 Function

4 GRASP + 4 ILS + 4 GA		
Results	with coop.	without coop.
Best solution	−1.0000	−0.9907
Standard Deviation	9.9889E-5	5.7999E-2
Average of the solutions	−0.9999	−0.9508
Best solution agent	ILS	ILS
% of global optimal solutions found	73.33	0
Exact Result	−1	

Furthermore, one can observe that with cooperation, the system converged to the optimum value of the function in 73.33 % of the executions, while in the experiment without cooperation, the system did not reach the optimal value in any of the executions.

In order to confirm that cooperation contributes for searching better solutions to optimization problems, two techniques of statistical analysis, box plot and ANOVA, were used.

Box plot is a nonparametric test that shows variation in samples of a statistical population without making any assumptions of the underlying statistical distribution. The spaces between the various parts of the box indicate the degree of dispersion and asymmetry in the data [17]. This method is useful to compare different sets of data regarding their homogeneity and trends. Thus, the results of the box plot obtained for a function minimization problem (XinSheYang03) can be seen in Fig. 4. The x-axis represents the two operating modes of the architecture, i.e., with and without cooperation. The y-axis, in turn, represents the best value of the objective function.

Figure 4 shows that the results with cooperation between the agents were more consistent, since the degree of dispersion was lower than that of the result obtained without cooperation. However, the results of the box plot analysis do not provide enough statistical evidence that there might be a significant difference between the data series, since the boxes shown in Fig. 4 exhibit an overlapping between them. Accordingly, after verified the premises of normality, randomness and homoscedasticity of the data set an analysis of variance (ANOVA) [17] was applied to the results found by the architecture to XinSheYang03 function.

Table 3 presents the results of ANOVA performed to XinSheYang03 function. ANOVA detects with a significance of 5 % that there are differences between the averages in the data series. This statement indicates that with cooperation between agents the result of the minimization problem of the XinSheYang03 function is improved. The guarantee of this test lies in the analysis of Table 3 since the value of P-value of 2.0313E-5 is lower than $\alpha = 0.05$, value adopted for the test. Therefore, there is statistical evidence, with a confidence interval of 95 %, that cooperation between agents enhances significantly the exploration of the search space of this type of optimization problem.

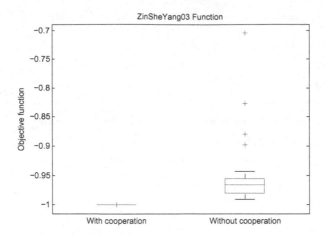

Fig. 4. Box Plot - XinSheYang03 function

Table 3. ANOVA - XinSheYang03 function

Source	SS	df	MS	F	Prob>F
Columns	0.03622	1	0.03622	21.53	2.0313E-5
Error	0.09755	58	0.00168		
Total	0.13377	59			

3.2   Experiment 2 - Knapsack Problem

The multidimensional 0–1 knapsack problem [8] is a variation of 0–1 knapsack problem. It is considered as a multidimensional problem because each object has a set of weights (features) that must meet the capabilities (restrictions) of the knapsack. Thus, if one or more weights of an object do not meet their respective capacity in the knapsack, then the object should not be part of the solution.

The instances used in the tests were taken from the file named mknap1 available in [3]. The architecture was executed with 16 agents with and without cooperation in 30 runs of 60 min each. The agents used were: 4 GRASP, 4 ILS, 4 SA and 4 GA. Table 4 shows the parametrization (also empirically established) of the metaheuristic agents used in this experiment.

The results of the average values obtained over 30 runs of the architecture with and without cooperation between agents are summarized in Table 5. As can be observed, the average of the values of the objective function obtained with the cooperation in all executions, indicates that the system converges toward a value closer to the optimum compared to the system without the cooperation. Furthermore, one can observe that with cooperation, the system converged to the optimal value of the function in 16.67 % of the executions, while in the experiment without cooperation, the system did not reach the optimal value in any of the executions.

Table 4. Metaheuristics Parameters - Knapsack Problem

Number of Agents	Metaheuristics	Solution modifers	Parameter values
4	GRASP	Local Search: Best Neighbour List of Candidates: Step	% of Randomness= 0.5 Nº of Iterations = 10
2	ILS	Local Search: Random Perturbation: Random	Perturbation Level = 7 Nº of Iterations = 200
2	ILS	Local Search: Best Neighbour Perturbation: Position	Perturbation Level = 10 Nº of Iterations = 100
2	GA	Crossover: Linear Combination Mutation: Random	Number of Parents = 2 Nº of Iterations = 100 Population Size = 30 Mutation Rate = 0.2 Crossover Rate = 0.7
2	GA	Crossover: Linear Combination Mutation: Position	Number of Parents = 2 No of Iterations = 200 Population Size = 20 Mutation Rate = 0.1 Crossover Rate = 0.8

Table 5. Experiment 2: knapsack Problem

4 GRASP + 4 ILS + 4 SA + 4 GA		
Results	with coop	without coop.
Best solution	16537	16521
Standard Deviation	12.3615	28.7419
Average of the solutions	16518.2333	16471.8000
Best solution agent	SA	ILS
% of global optimal solutions found	16.67	0
Literature Result	16537	

Following the procedure carried out in the previous experiment and to confirm that cooperation contributes effectively for searching better solutions to optimization problems, the techniques of statistical analysis, box plot and ANOVA were also applied in this experiment.

Thus, the box plots of architecture with and without cooperation were initially generated, as shown in Fig. 5. In this figure, it can be seen that for this maximization problem (knapsack), the results with cooperation were more consistent, since its degree of dispersion was lower than that of the simulation without cooperation. In addition, it is clear that there is a significant statistical difference between the data series, since there was not overlapping of the boxes.

In order to confirm this hypothesis, ANOVA was also applied to the results found. The analysis suggests that there are differences in the use of cooperation

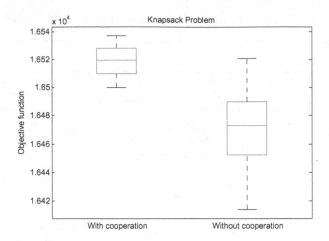

Fig. 5. Box Plot - Knapsack problem

Table 6. ANOVA - Knapsack Problem

Source	SS	df	MS	F	Prob>F
Columns	32340.8	1	32340.8	66.08	3.7285E-11
Error	28388.2	58	489.5		
Total	60729	59			

between agents due to the fact that P-Value = 3.7285E-11 was lower than α = 0.05 (Table 6). ANOVA shows that cooperation amongst agents improves the quality of results of the Bimasco architecture for searching better solutions to optimization problems with a confidence interval of 95 %.

4   Conclusion

This work proposes a new architecture named Bimasco that consists of an bioinspired multi-agent autonomous system for solving optimization problems of different classes. This system models the dynamics of interactions among agents and regions, so that it operates as a self-organizing discrete dynamical system.

This work proposes two preliminary computational experiments of different classes, including one problem involving an analytical function and the knapsack problem. The results show that the dynamics of the implemented model is most effective when there is cooperation between the agents. Cooperation creates a wide range of solutions and thereby increases the likelihood of finding good solutions to various classes of optimization problems. From the experiments, it was observed that cooperation in a multi-player environment for solving optimization problems, can make the system more efficient and effective in exploring the search space of the problem. Consequently, in order to validate our conclusions further experiments should be carried out on a wider range of problems, such as, a large set of benchmark continuous functions and a large set of knapsack benchmark problems.

Finally, the architecture is currently being extended in order to solve multi-objective problems. In addition, new generalized metaheuristic agents, constructive methods, local search strategies and population-based search are being implemented in the architecture, aiming at increasing the "biodiversity" of agents, thereby improving the quality of the solutions obtained by the system.

Acknowledgments. The authors thank the support of CAPES-Brazil under Procs. BEX 7354/14-2, CNPq-Brazil, FAPEMIG and CEFET-MG.

References

1. Aydin, M.E.: Collaboration of heterogeneous metaheuristic agents. In: Proceedings, International Conference on Digital Information Management (ICDIM), pp. 540–545. IEEE, Thunder Bay (2010)
2. Barbulescu, L., Watson, J., Whitley, L.D.: Dynamic representations and escaping local optima: Improving genetic algorithms and local search. In: Proceedings of the Seventeenth National Conference on Artificial Intelligence, July 30 -August 3, 2000, Austin, Texas, USA, pp. 879–884. AAAI Press, California (2000). http://www.aaai.org/Library/AAAI/2000/aaai00-135.php
3. Beasley, J.E.: Or-library: Multidimensional knapsack problem (1990). http://people.brunel.ac.uk/mastjjb/jeb/orlib/mknapinfo.html
4. Crain, T., Bishop, R.H., Fowler, W., Rock, K.: Interplanetary flyby mission optimization using a hybrid global-local search method. J. Spacecraft Rockets **37**(4), 468–474 (2000)
5. Dorigo, M., Birattari, M., Stützle, T.: Ant colony optimization. IEEE Comput. Intell. Mag. **1**(4), 28–39 (2006)
6. Feo, T.A., Resende, M.G.: Greedy randomized adaptive search procedure. J. Global Optim. **6**(2), 109–133 (1995)
7. Fernandes, F., de Souza, S., Silva, M., Borges, H., Ribeiro, F.: A multiagent architecture for solving combinatorial optimization problems through metaheuristics. In: Proceedings of the IEEE International Conference on Systems, Man and Cybernetics, SMC 2009, pp. 3071–3076, October 2009
8. Fréville, A.: The multidimensional 0–1 knapsack problem: an overview. Eur. J. Oper. Res. **155**(1), 1–21 (2004)
9. Gavana, A.: Global optimization test functions index (2013). http://infinity77.net/global_optimization/test_functions.htm
10. Holland, J.H., Reitman, J.S.: Cognitive systems based on adaptive algorithms. ACM SIGART Bull. **63**, 49–49 (1977)
11. Jin, J., Crainic, T.G., Løkketangen, A.: A cooperative parallel metaheuristic for the capacitated vehicle routing problem. Comput. Oper. Res. **44**(1), 33–41 (2014)
12. Johnson, S.: Emergence: The Connected Lives of Ants, Brains, Cities, and Software, 1st edn. Simon and Schuster, Nova York, NY (2002)
13. Kirkpatrick, S.: Optimization by simulated annealing: quantitative studies. J. Stat. Phy. **34**(5–6), 975–986 (1984)

14. Lourenço, H.R., Martin, O.C., Stützle, T.: Iterated local search. Sci. Kluwer **57**, 321–353 (2002)
15. Løvbjerg, M., Rasmussen, T.: Hybrid particle swarm optimiser with breeding and subpopulations. In: Proceedings of the third Genetic and Evolutionary Computation Conference (GECCO-2001) (2001)
16. Milano, M., Roli, A.: Magma: a multiagent architecture for metaheuristics. IEEE Trans. Syst. Man Cybern. **34**(2), 925–941 (2004)
17. Montgomery, D.C.: Design and Analysis of Experiments. Wiley, Hoboken (2008)
18. Preux, P., Talbi, E.G.: Towards hybrid evolutionary algorithms. Int. Trans. Oper. Res. **6**(6), 557–570 (1999)
19. de Souza, D.: Generalisation of metaheuristic software agents. Master's thesis, Belo Horizonte (2014). http://www.lsi.cefetmg.br/Bimasco
20. Talbi, E.G.: A taxonomy of hybrid metaheuristics. J. Heuristics **8**(5), 541–564 (2002)
21. Vicini, A., Quagliarella, D.: Airfoil and wing design through hybrid optimization strategies. AIAA J. **37**(5), 634–641 (1999)
22. Wolpert, D.H., Macready, W.G.: No free lunch theorems for optimization. IEEE Trans. Evol. Comput. **1**(1), 67–82 (1997)
23. Xie, X.F., Liu, J.: Multiagent optimization system for solving the traveling salesman problem (TSP). IEEE Trans. Syst. Man Cybern. **39**(2), 489–502 (2009)
24. Xu, P.: A hybrid global optimization method: the multi-dimensional case. J. Comput. Appl. Math. **155**(2), 423–446 (2003)

Novelty-Driven Particle Swarm Optimization

Diana F. Galvao[1], Joel Lehman[2], and Paulo Urbano[1($\boxtimes$)]

[1] Faculty of Sciences, BioISI Biosystems and Integrative Sciences Institute,
University of Lisboa, Campo Grande, Lisboa, Portugal
`fc37298@alunos.fc.ul.pt`, `pub@di.fc.ul.pt`
[2] IT University of Copenhagen, Copenhagen, Denmark
`jleh@itu.dk`

Abstract. Particle Swarm Optimization (PSO) is a well-known population-based optimization algorithm. Most often it is applied to optimize objective-based fitness functions that reward progress towards a desired objective or behavior. As a result, search increasingly focuses on higher-fitness areas. However, in problems with many local optima, such focus often leads to premature convergence that precludes reaching the intended objective. To remedy this problem in certain types of domains, this paper introduces Novelty-driven Particle Swarm Optimization (NdPSO), which is motivated by the novelty search algorithm in evolutionary computation. In this method particles are driven only towards instances significantly different from those found before. By ignoring the objective this way, NdPSO can circumvent the problem of deceptive local optima. Because novelty search has previously shown potential for solving tasks in genetic programming, this paper implements NdPSO as an extension of the grammatical swarm method, which combines PSO with genetic programming. The resulting NdPSO implementation is tested in three different domains representative of those in which it might provide advantage over objective-driven PSO. That is, deceptive domains in which it is easy to derive a meaningful high-level description of novel behavior. In each of the tested domains NdPSO outperforms both objective-based PSO and random-search, demonstrating its promise as a tool for solving deceptive problems.

Keywords: Particle Swarm Optimization · Novelty search · Grammatical evolution · Grammatical swarm · Deceptive problems

1 Introduction

Particle Swarm Optimization (PSO) is a biologically-inspired population-based optimization algorithm [1]. Although PSO is a popular and effective algorithm, like other population-based methods it is susceptible to converge prematurely to local optima when applied to complex or deceptive problems [2,3]. Most applications of PSO optimize an objective-based fitness function which estimates the progress to the desired outcome, e.g. minimizing squared error or heuristic similarity to a goal behavior. Guiding the search directly towards the ultimate goal

© Springer International Publishing Switzerland 2016
S. Bonnevay et al. (Eds.): EA 2015, LNCS 9554, pp. 177–190, 2016.
DOI: 10.1007/978-3-319-31471-6_14

causes increasing focus on higher-fitness areas at the expense of lower-fitness ones, reducing the overall exploration of the search space. In simple problems this focus aids efficiency, but can be harmful in deceptive ones. That is, the pervasiveness of local optima may render the objective accessible only by significant travel through areas with low objective-based fitness. By pruning such areas from consideration, an objective-driven algorithm may be unlikely to reach the desired objective.

Because local optima are a well-known issue in search, many researchers have proposed variations of PSO to circumvent premature convergence [4,5]. While such variations may outperform the standard PSO algorithm in domains with limited deception, because they remain guided by heuristic distance to the objective, they are still vulnerable to premature convergence if the objective function is sufficiently deceptive. In this way, there is a clear relationship between objective-based search and premature convergence. Thus to avoid premature convergence in very deceptive domains it may paradoxically be necessary to guide search without considering the ultimate objective.

Novelty search is an evolutionary algorithm (EA) which takes this radical step [6], and has successfully been applied in neuroevolution [6,7] and genetic programming [8,9]. The core insight motivating novelty search is that novelty, i.e. demonstrating qualitative difference from previously encountered individuals, is a valuable source of information. Thus instead of guiding search by estimated distance to the search's objective, novelty search is driven towards instances significantly different from those found before. By ignoring the objective completely, novelty search circumvents the problem of deception inherent in objective-based algorithms. Of course, without any pressure to optimize towards the objective, intuitively a raw search for novelty may seem unlikely to be effective at solving problems. Yet if measures of novelty are constructed that capture important dimensions of behavioral variation in the domain, the surprising result is a practical algorithm for solving deceptive problems [6,8,10]. The insight is that often demonstrating novelty requires exploiting meaningful regularities in a domain.

A key motivation for this paper is that because PSO is also susceptible to deception, integrating a drive towards novelty might sometimes also benefit the effectiveness of PSO. This paper thus introduces Novelty-driven PSO (NdPSO), a tool to combat the pathology of premature convergence in PSO. Because novelty search has shown prior promise in combination with genetic programming [10], here NdPSO is implemented as an extension of the Grammatical Swarm (GS) method [11], which is a PSO-based version of a GP technique called Grammatical Evolution (GE). This implementation is thus called novelty-driven Grammatical Swarm (NdGS).

Experiments in this paper test NdGS in three domains representative of those for which the algorithm might be most effective. Such domains are deceptive (otherwise an objective-based search method is likely to be more effective) and provide an intuitive way to characterize a space of behaviors that captures important domain features (otherwise it is difficult to quantify novelty). The first domain requires evolving a hidden sequence obscured by a deceptive fitness function. The other domains are reinforcement learning benchmarks imported from genetic programming that are also known to be deceptive. The experiments compare the performance of NdGS, traditional objective-driven GS and purely random search. Across the tested domains NdPSO performs the best, highlighting its potential for solving deceptive problems.

2   Background

The next sections describe the PSO algorithm used in the experiments, the Grammatical Swarm extension that enables PSO to evolve programs, and the novelty search method merged with PSO in this paper's approach.

2.1   Particle Swarm Optimization

This section reviews the main concepts of PSO, a population-based optimization algorithm inspired by schooling and flocking behaviors of animals, introduced by Kennedy and Eberhard [12].

In PSO, the population (or swarm) is composed of particles moving through a $\mathbb{R}^d$ search space, that optimize a fitness function with the following domain $f : \mathbb{R}^d \to \mathbb{R}$, with d representing the dimensionality of the search space.

Each particle $i \in (1, 2, 3, ..., N)$ is associated with two d-dimensional vectors, one recording its position x and the other its velocity v.

Particles also store a summary of their previous experiences in a simple memory component. In particular, the particle records the position of the maximum fitness value it has encountered, called its personal-best or *pbest*. Particles also share information with each other, and record the point in the search space where the overall best fitness has been obtained among all particles. This point, which a particle may not have visited (but has heard from another particle) is called its global-best *gbest*. These components help balance exploiting promising areas with exploring more broadly [13].

In practice, communication between particles is often restricted by use of a neighborhood topology, meaning that a particle's *gbest* may be calculated from the best search locations recorded by its neighboring particles [14]. The most commonly chosen topology is a fully-connected neighborhood [14]. In such a topology all particles directly communicate with all other particles. Thus information propagates quickly, which can cause fast convergence. In the Ring topology, particles communicate only with their immediate neighbors based on their position in a circular list, causing information to flow more slowly than in the Fully-connected topology. The result is that some groups of particles can converge to one point in the search space, while other groups can converge to others. The Von Neumann topology offers an intermediate speed of information flow; each particle is assigned a fixed location on a two-dimensional toroidal neighborhood grid, and can communicate with its nearest neighbors in the four cardinal directions.

At each time step t, the velocity of each particle is adjusted as a function of its position, previous velocity, *pbest* and *gbest*. In particular, the ith particle's velocity is updated as follows:

$$v_i(t+1) = \omega.v_i(t) + \varphi_1.r_1(pbest_i(t) - x_i(t)) + \varphi_2.r_2(gbest(t) - x_i(t)), \qquad (1)$$

where ω is a parameter specifying the particle's inertia, which determines how strongly the particle maintains its previous velocity (i.e. the higher the inertia, the slower that velocity changes). The experiments apply a dynamic inertia value, which is initialized to 0.9 and during search decreases linearly to 0.4 using Eq. 3. The real numbers r_1 and r_2 are chosen randomly within an interval (typically between 0 and 1), and φ_1 and φ_2 are the acceleration coefficients. It is also common to restrict the maximum velocity ($v_{max} \in [-v_{max}, v_{max}]$) to prevent instability. Note that the particle's maximum velocity may be static, as in our experiments, or calculated dynamically [15]. The particle's new position is calculated according to Eq. 2.

$$x_i(t+1) = x_i(t) + v_i(t+1). \qquad (2)$$

$$\omega = \omega_{max} - \frac{\omega_{min}}{maxIterations} * NumIteration, \qquad (3)$$

In this way, the particles are driven through the search space towards locally optimal points. Because they are attracted both to their own best position and the overall best position, over time a consensus may emerge as knowledge of the seemingly most promising point in the search space spreads through all neighborhoods. This process often results in convergence.

Barebones PSO. The Barebones PSO algorithm was developed by Kennedy [5], and it is a variant of the standard PSO algorithm designed to mitigate premature convergence. The main idea is to minimize the degree to which good performance depends upon well-tuned settings of parameters like ω, φ_1 and φ_2. While other PSO variations have similar motivation, Barebones PSO completely eliminates these parameters from the algorithm. The main difference from the standard PSO algorithm is a simplified position update (Eq. 4) where $\sigma = |pbest_{ij}(t) - gbest_j(t)|$.

$$x_{ij}(t+1) \sim N \left(\frac{pbest_{ij}(t) + gbest_j(t)}{2}, \sigma \right), \qquad (4)$$

As shown above, the position of each particle is iteratively sampled from the Gaussian distribution. Barebones PSO favors exploration in the earlier stages of the simulation because the personal best positions will at first be far from the global best one, leading to sampling from a probability distribution with a higher variance. As the simulation proceeds and personal bests converge to the global best, variance will approach zero, thereby focusing on exploitation.

Barebones PSO is a popular variation of the standard PSO algorithm because it has few parameters, and because of its aim to deal with the problem of premature convergence. Over many studies, Barebones PSO has provided better results than the standard PSO algorithm [5,16,17]. These factors motivate its use as an additional comparison algorithm in the this paper's experiments.

2.2 Grammatical Swarm

The Grammatical Swarm (GS) algorithm [11] combines PSO with the Grammatical Evolution (GE) mapping process. GE is an Evolutionary Algorithm (EA) able to evolve computer programs in any language that can be described in grammatical form [18]. The insight is that programs can be represented based on a Backus Naur Form (BNF) grammar instead of on parse trees as in tree-based GP.

In GS, most commonly particles are given fixed-length dimensionality, which is the approach adopted in these experiments. To map from PSO's floating point representation to the integer codon representation of GE, each real-valued element of the particle's position is rounded to the nearest integer value. This new array of integers is then mapped through a fixed grammar into a program.

Mapping Process. In the GS mapping process a particle's location in the search space is processed to construct a program (which can then be evaluated in the domain).

An example of the mapping process in GE and GS is shown in Fig. 1. Note that an integer representation is used in this example for simplicity of understating. In this example, the first codon (00001110) is converted to its decimal value (14). From the underlying grammar, the start symbol <E> has two possible rules that can be applied. The modulus of the value of the codon with the number of alternatives determines which rule is selected ($Alternative = CodonValue \% Num. of Alternatives$).

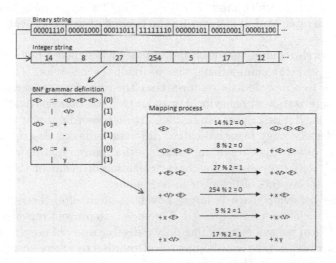

Fig. 1. GE and GS mapping process. This example is based on [11].

For this example, 14 % 2 = 0, which means that the first rule, <O> <E> <E>, is selected, i.e. the start symbol is replaced with those three non-terminal symbols. Next, the leftmost symbol <O> and the second codon (8) are processed in the same way. The process then iterates until no non-terminal symbols remain. When the genotype has fewer codons than non-terminals, the genome is "wrapped", i.e. the algorithm continues reading again from the beginning of the genotype. The number of times the genotype wraps is limited in practice to avoid infinite cycles. In this case, the respective particle is marked as invalid. Our preliminary experiments revealed that if particles are forced to be valid (i.e. by updating a particle's velocity until its position becomes valid), the performance of all algorithms improved considerably. Therefore, all results in this paper include this procedure.

2.3 Novelty Search

The novelty search algorithm was introduced by Lehman and Stanley in 2008 as an alternative to objective-based optimization in evolutionary computation [6]. The key idea behind novelty search is to ignore the objective of the search, and instead reward novel individuals, i.e. those with behaviors different from those previously encountered in the search. The insight is that by not optimizing a measure of progress towards the objective, novelty search is not susceptible to premature convergence to local optima. Of course, the significant trade-off is that the search as a whole becomes less explicitly controlled.

While novelty search may record the value of the underlying objective-based fitness function to track whether any solutions have been discovered, this fitness function does not guide the search. Instead novelty search applies a novelty metric, which measures the novelty of an individual in comparison to other individuals in the search, according to its behavior. This enables rewarding individuals with novel behaviors, which incentivizes exploring the space of behaviors. This exploration can often lead to discovering the desired objective of search as a side effect.

Novelty search requires that each individual be assigned a behavior descriptor, which is a vector summarizing the individual's behavior. Thus applying novelty search to a new domain requires that the experimenter devises a quantitative characterization of behavior. For example, in a maze navigation task an individual's behavior descriptor can be its Cartesian coordinates at the end of an evaluation. Because they bias the search, different behavior descriptors applied to the same domain may result in varied performance. In this way, effective novelty search may depend upon a characterization of behavior that succinctly captures relevant aspects of behavior in a domain.

When the behavior space is large, novelty search often benefits when an archive of past behaviors is maintained. The idea is to prevent repeatedly cycling through a series of behaviors, reflecting only a fleeting sense of novelty. By archiving past behaviors, novelty can be measured relative to where search has been and where it currently is, thus incentivizing behaviors that are genuinely novel. Two common archiving strategies are to add behaviors with novelty that is higher than a threshold value, or to select individuals randomly to archive. In the experiments presented here, all behaviors had a five percent probability of being archived.

To calculate the novelty of an individual requires defining the distance between its behavior descriptor and that of others in the population and in the archive. Given the behavior descriptor and the distance metric between such descriptors, the novelty score is calculated as shown in Eq. 5, where μ_i is the ith nearest neighbor of x, and $dist$ is a Euclidean distance metric.

$$\rho(x) = \frac{1}{k} \sum_{i=0}^{k} dist(x, \mu_i). \tag{5}$$

The novelty of each individual is the average distance of its behavior to its k-nearest neighbors. This way, individuals in a less dense area of the behavior space are given higher novelty scores, thus creating pressure in the search towards further novelty.

3   Novelty-Driven PSO

Algorithm 1, below, provides the pseudo-code for the novelty-driven PSO algorithm. Introducing novelty search into PSO does not change the core of the algorithm (only in the reward scheme driving search), resulting in pseudocode that is not significantly different from the standard PSO algorithm.

The main change to PSO is to replace the objective-based fitness function with a novelty metric. The standard velocity update equation (Eq. 1) shows that particles are attracted by the best position they have encountered and the overall

---

Algorithm 1. Novelty-driven PSO algorithm

```
 1: procedure NSPSOPROCEDURE
 2: createPopulation: foreachParticle
 3:     setRandomPosition
 4:     setRandomVelocity
 5:     evaluatePopulation
 6:     setPbestCurrent
 7:     executeNovelty
 8:     setPnovelCurrent
 9:     end foreach
10:     updateGbest
11:     updateGnovel
12: loop:
13:     if NotDone then foreachParticle
14:         calculateVelocity
15:         calculateNewPosition
16:         evaluatePopulation
17:         updatePbest
18:         executeNovelty
19:         updatePnovel
20:         end foreach
21:         updateGbest
22:         updateGnovel
23:     close;
```

---

best position. In novelty-driven PSO, the equation is the same, but the concept of "best position" changes. That is, in novelty-driven PSO, the *pbest* position is not the position where the particle obtained best fitness but the position where it has encountered highest novelty. Similarly, the overall best position (*gbest*) is the position where the maximum value of novelty has been obtained. For this reason, these quantities are referred to as *pnovel* and *gnovel*.

One important aspect of novelty is that it is both relative and dynamic. That is, each particle calculates its novelty with respect to other particles' current behaviors and past archived behaviors. In this way, a highly novel behavior becomes less novel over time as other particles become drawn to it, and it is added to the archive. For this reason, the tracking of the *pnovel* and *gnovel* take this dynamism into account. In particular, the novelty of each particle's *pnovel* behavior is recalculated each iteration. If the novelty score of the current particle position is higher than the recalculated novelty score of its *pnovel* than the latter is updated. In this case, the current position will overwrite *pnovel*, and the particle's behavior will also be cached as the *pnovel* behavior. *Gnovel* is the *pnovel* over all particles with the highest score. Importantly, because the behavior vector for each *pnovel* is cached, no additional domain evaluations are required for such recalculations, and they thus incur little computational overhead.

Note that when more than one *pnovel* shares the same highest novelty score, and the previous *gnovel* is among such high-scoring *pnovel* search points, empirically it was found to be important to conserve the current *gnovel*. That is, if the current *gnovel* particle is one the top ranked *pnovel* particles, it always is chosen to remain as the *gnovel*. But if the current *gnovel* is not among of the tied top *pnovel* scores, one of the top ranked *pnovel* is randomly chosen. That is, *gnovel* changes only when there a *pnovel* with a strictly higher novelty score.

4   Experiments

The aim of these experiments is to compare the performance of the traditional objective-driven GS method with the performance of the NdPSO algorithm proposed in this paper. Recall that novelty search pursues novel behaviors, which makes it best-suited for deceptive problems in which it is possible and intuitive to characterize an individual's behavior. Thus the choice of test domains reflects the type of problem for which the approach is likely to be appropriate. Objective-driven GS and NdGS are tested in three domains: Mastermind, the Santa Fe trail, and a Maze navigation problem.

All experiments have been performed in Netlogo, using a Java implementation of the GE algorithm [19] (the jGE library) and the respective Netlogo extension [20].

4.1   Mastermind

Inspired by the classic board game, in the Mastermind domain considered here the task is to discover the correct sequence before the maximum number of

<pin> ::= <pin> <pin> | 0 | 1 | 2 | 3

Fig. 2. BNF-Koza grammar definition for the Mastermind problem.

attempts is exhausted. As in O'Neill [11], four possible colors are considered, and the correct code for the eight given positions was fixed to 3 2 1 3 1 3 2 0. NdGS and GS both use the same grammar as in the original GS experiment (Fig. 2).

In this domain, objective-based fitness is scored as follows: One point is awarded for each correctly colored pin (regardless of its position, although limited in extent by the total number of pins in the target sequence with that color). If all pins are in the correct position, an additional point is awarded. If the genotype has more than eight codons, it is truncated to the first eight. The fitness score is normalized by dividing the raw score by the maximum score possible ($fitness = score/maxScore$). Note that $maxScore$ in this experiment is 9. By design, this problem is highly deceptive, with local optima corresponding to all sets of correct colors in wrong positions.

In contrast to objective-based PSO, novelty-driven PSO requires a characterization of behavior. In this problem two distinct characterizations were considered: behaviorMm1 - the fitness value, and a behavior inspired by the original game: behaviorMm2 - a tuple of two integers consisting of the number of correct colors and the number of correct positions. Two things are important to note. First, searching for novel values of objective fitness is different than simple objective-based search. That is, novelty search will be driven to accumulate all possible fitness values, not only the largest ones. Thus the performance of objective-driven GS and novelty-driven GS may diverge even though they are driven by the same underlying information. Second, behaviorMm2 provides an ideal decomposition of the domain, one which is unavailable to the objective fitness function (i.e. both colors and placements are important). Thus this characterization highlights the potential for injecting experimenter knowledge into the search process, which may otherwise be complicated in objective-driven search because of the need to reduce performance information to a single number.

4.2   Santa Fe Trail

The Santa Fe trail is a difficult and popular benchmark in both GP [21] and GE [18].

The goal is to evolve a computer program that can efficiently guide an artificial ant to eat all pieces of food placed in the trail (Fig. 3). Beginning from the upper left corner, the artificial ant can move forward in the direction it is currently facing or turn 90 degrees to the right or left. Each action takes one discrete unit of time to perform. The ant can perceive if the cell in front of it contains food, an operator that executes instantaneously (i.e. it does not consume any time). Figure 4 shows the grammar used in this domain.

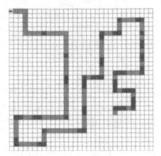

Fig. 3. The Santa Fe Trail.

```
<code> ::= <line> | <code> <line>
<line> ::= <condition> | <op>
<condition> ::= ifelse food-ahead [ <line> ][ <line> ]
<op> ::= turn-left | turn-right | move
```

Fig. 4. BNF-O'Neill grammar definition for the Santa Fe trail.

For objective-driven GS, the traditional fitness function simply counts the units of food eaten by the ant after all time has been exhausted. The standard maximum number of steps is used in this experiment, i.e. 615.

For novelty-driven GS, two behavior descriptors are applied. A simpler descriptor - behaviorSF1 - adopts the fitness function as the characterization of behavior, as in the Mastermind domain. A more informative characterization - behaviorSF2 - considers the amount of food eaten, with the constraint that the eaten units must not be disconnected from other eaten units along the length of the trail.

For example, if the ant first eats 3 food units that follow the trail, then leaves the trail and eats one more unit in another area of the trail, its behavior descriptor is appended with a 3 (although the ant ate in total 4 units of food), because the last unit is not connected to any other eaten units along the true path of the trail. However, if the ant eats 3 food units at the beginning of the trail, goes off the trail, and collects three more units along a later part of the trail, the score will then be 6. Additionally, this second characterization is sampled over time to provide temporal information about the ant's behavior. In particular, it is sampled every 41 timesteps, resulting in a vector of length 15 by the completion of an ant's evaluation.

4.3 Maze Navigation Problem

The Medium Map domain (Fig. 5) is a deceptive and discrete maze navigation task introduced by Lehman and Stanley [22]. The goal in this domain is to find a program that guides an agent in a grid-world domain to the goal location before exhausting the time limit. In our experiments the time limit was set to 500 steps. This maze is suitable for testing GS and NdGS because the placement of the walls create deception. That is, the shortest path to the goal is blocked, meaning that to solve the task requires exploring areas that superficially appear further from the goal.

The possible actions for each agent are: *turn-left, turn-right, move* and the boolean operators *wall-ahead, wall-left* and *wall-right* (Fig. 6). Similarly to as in

Fig. 5. Medium Map used for the maze navigation problem.

```
<expr> ::= <line> | <expr> <line>
<line> ::= ifelse <condition> [ <expr> ][ <expr> ] | <op>
<condition> ::= wall-ahead? | wall-left? | wall-right?
<op> ::= turn-left | turn-right | move
```

Fig. 6. BNF-Koza grammar definition for the Medium Maze problem.

the Sante Fe Trail, the turning actions respectively turn the agent 90 degrees left or right, while the move operation causes the agent to progress forward one unit in the direction it currently faces. All but the last three actions consume a timestep.

Defining *dist* as the euclidean distance of the agent from the goal location, the fitness function for objective-based GS is calculated as $fitness = \frac{1}{1+dist}$.

For novelty-driven GS, the behavior descriptor adopted was the coordinates of the agent's ending position - behaviorMP1. In this way, objective-based search is looking for ways to get closer to the goal, while novelty search instead explores how to reach a diversity of places within the maze.

4.4   Experimental Parameters

As in O'Neill's original GS experiments [11], the following parameter settings were adopted in all experiments: $\varphi_1 = \varphi_2 = 1.0$, $\omega_{max} = 0.9$, $\omega_{min} = 0.4$ and $v_{ij} \in [-255, 255]$. A maximum of 10 wraps was allotted for an individual to be considered valid, and a coordinate's value was bounded between 0 and 255, inclusive. When a particle exceeds the maximum or minimum value for a particular dimension, the value is clipped such that it lies on the extreme of that range. The swarm consisted of 30 particles, with a search space of dimensionality 100. Simulations ran for 1000 iterations.

Three different neighborhood topologies are tested: Fully-connected, Ring and Von Neumann. All results reported for objective-driven GS and GS (Barebones) use the Fully-connected topology which provided the best results. Interestingly, with objective-driven GS the Ring and Von Neumann topologies have effect opposite from when used in NdGS (where their use improves results significantly).

In the Santa Fe trail and in the maze navigation problem 3 nearest neighbors were used for calculating novelty in NdGS, and 15 nearest neighbors were used in the Mastermind problem. Because the behavior *NdGS-behaviorSF2* in Santa Fe is composed of many samples taken over an ant's evaluation, the best result was obtained by including the archive.

5   Results and Discussion

The results presented in Table 1 are the best obtained for each algorithm in the considered domain, over the combinations of topologies and other parameters described in the section above. In all problems NdPSO outperforms the objective-based algorithms tested, both in number of solutions found and in the best fitnesses discovered averaged over all runs. In particular, the average best fitnesses are significantly higher for the NdPSO methods than for any of the control algorithms (Student's t-test; $p < 0.05$). Interestingly, in the Mastermind problem, which is considered a difficult GP problem, NdPSO managed to succeed in all runs while objective-driven GS achieved success in only 14 % of runs.

While every NdPSO treatment outperforms the control algorithms, it is important to note that the selection of the behavior descriptor is a significant factor. In particular, a domain-specific descriptor that integrates additional information than the raw objective-based fitness measure led to further performance gains. Additionally, performance of NdPSO is affected by interactions between the behavior characterization and whether or not an archive is included. In particular, when using behaviorSF2 in the Santa Fe Trail and the Maze navigation problem, an archive provided better results.

Table 1. Comparison of the results obtained for Grammatical Swarm, Novelty-driven Grammatical Swarm and Random Search averaged over 100 runs

	Mean best fitness	Std Deviation	Median	Successful runs
Mastermind				
GS	0.90	0.04	0.89	14
GS (Barebones)	0.87	0.04	0.89	2
NdGS-behaviorMm1	0.92	0.05	0.89	25
NdGS-behaviorMm2	**1.00**	**0.00**	**1**	**100**
Random	0.86	0.05	0.89	0
Santa Fe Trail				
GS	78.31	15.65	89	52
GS (Barebones)	74.25	16.95	88	49
NdGS-behaviorSF1	85.46	8.54	89	75
NdGS-behaviorSF2	**87.89**	**6.52**	**89**	**78**
Random	75.81	15.74	88	40
Maze problem				
GS	0.43	0.36	0.24	27
GS (Barebones)	0.50	0.36	0.31	34
NdGS-behaviorMP1	**0.68**	**0.38**	**1**	**58**
Random	0.30	0.30	0.17	15

Despite its promise in previous work, applying objective-based Barebones GS did not improve upon on the results of standard GS, performing worse than GS in 2 out of 3 domains. This failure may be particular to the chosen domains, or may result from combining the Barebones approach with GS.

A final interesting result is that the best neighborhood topology differs across objective-driven and novelty-driven GS. As described earlier, the best performance for objective-driven GS (with and without Barebones PSO) resulted from the fully-connected topology, whereas in NdGS the Ring topology substantially improved performance. This result highlights a qualitative difference between an objective-driven search and a novelty-driven one; best practices for one paradigm may not directly transfer to the other.

6 Conclusion and Future Work

This paper introduced Novelty-driven PSO (NdPSO), a method that aims to mitigate the challenge of premature convergence in traditional objective-driven PSO. NdPSO was combined with Grammatical Swarming and tested in three domains. In each domain it outperformed two objective-based control algorithms and random search. In this way, NdPSO shows promise for solving deceptive PSO problems and encourages further follow-up investigation. In particular, future research will compare NdPSO to novelty-driven GE, to examine how its performance compares to the evolutionary novelty search that inspired it.

Acknowledgments. This work was funded by FCT project EXPL/EEI-SI/1861/2013.

References

1. Kennedy, J., Eberhart, R.: Particle swarm optimization. In: Proceedings of IEEE International Conference on Neural Networks, vol. 4, IEEE (1995)
2. Peer, E.S., Van den Bergh, F., Engelbrecht, A.P.: Using neighbourhoods with the guaranteed convergence PSO. In: Proceedings of the 2003 IEEE Swarm Intelligence Symposium, SIS 2003, pp. 235–242. IEEE (2003)
3. Ratnaweera, A., Halgamuge, S.K., Watson, H.C.: Self-organizing hierarchical particle swarm optimizer with time-varying acceleration coefficients. IEEE Trans. Evol. Comput. **8**(3), 240–255 (2004)
4. Vesterstrom, J., Thomsen, R.: A comparative study of differential evolution, particle swarm optimization, and evolutionary algorithms on numerical benchmark problems. In: Congress on Evolutionary Computation, CEC2004, vol. 2. IEEE (2004)
5. Kennedy, J.: Bare bones particle swarms. In: Proceedings of the 2003 IEEE Swarm Intelligence Symposium, SIS03, pp. 80–87. IEEE (2003)
6. Lehman, J., Stanley, K.O.: Exploiting open-endedness to solve problems through the search for novelty. In: Proceedings of the Eleventh International Conference on Artificial Life (ALIFE XI). MIT Press, Cambridge (2008)

7. Lehman, J., Stanley, K.O., Miikkulainen, R.: Effective diversity maintenance in deceptive domains. In: Proceeding of the Fifteenth Annual Conference on Genetic and Evolutionary Computation Conference, pp. 215–222. ACM (2013)
8. Lehman, J., Stanley, K.O.: Efficiently evolving programs through the search for novelty. In: Proceedings of the 12th Annual Conference on Genetic and Evolutionary Computation, pp. 837–844. ACM (2010)
9. Lehman, J., Stanley, K.O.: Novelty search and the problem with objectives. In: Riolo, R., Vladislavleva, E., Moore, J.H. (eds.) Genetic Programming Theory and Practice IX, pp. 37–56. Springer, New York (2011)
10. Urbano, P., Georgiou, L.: Improving grammatical evolution in santa fe trail using novelty search. In: Advances in Artificial Life, ECAL, vol. 12 (2013)
11. ONeill, M., Brabazon, A.: Grammatical swarm: The generation of programs by social programming. Natural Comput. 5(4), 443–462 (2006)
12. Kennedy, J., Eberhart, R.: Particle swarm optimization. In: Proceedings of IEEE International Conference on Neural Networks, vol. 4 (1995)
13. Ozcan, E., Mohan, C.K.: Analysis of a simple particle swarm optimization system. Intell. Eng. Syst. Through Artif. Neural Netw. 8, 253–258 (1998)
14. Medina, A.J.R., Pulido, G.T., Ramrez-Torres, J.G.: A comparative study of neighborhood topologies for particle swarm optimizers. In: IJCCI, pp. 152–159 (2009)
15. Ali, M., Kaelo, P.: Improved particle swarm algorithms for global optimization. Appl. Math. Comput. 196, 578–593 (2008)
16. Omran, M., Al-Sharhan, S.: Barebones particle swarm methods for unsupervised image classification. In: IEEE Congress on Evolutionary Computation, CEC 2007, IEEE (2007)
17. Yao, J., Han, D.: Improved barebones particle swarm optimization with neighborhood search and its application on ship design. Math. Prob. Eng. 2013 (2013)
18. Ryan, C., Collins, J.J., Neill, M.O.: Grammatical evolution: evolving programs for an arbitrary language. In: Banzhaf, W., Poli, R., Schoenauer, M., Fogarty, T.C. (eds.) EuroGP 1998. LNCS, vol. 1391, p. 83. Springer, Heidelberg (1998)
19. Georgiou, L., Teahan, W.J.: jGE-A Java implementation of Grammatical Evolution. In: Proceedings of the 10th WSEAS International Conference on SYSTEMS, Vouliagmeni. Athens: WSEAS (2006)
20. Georgiou, L., Teahan, W.J.: Grammatical evolution and the santa fe trail problem. In: Evolutionary Computation (ICEC 2010) (2010)
21. Koza, J.R.: Genetic evolution and co-evolution of computer programs. In: Langton, C.T.C., Farmer, J.D., Ras-mussen, S. (eds.) Artificial Life II. SFI Studies in the Sciences of Complexity, vol. X, pp. 603–629. Addison-Wesley, Santa Fe (1991)
22. Lehman, J., Stanley, K.O.: Efficiently evolving programs through the search for novelty. In: Proceedings of the 12th Annual Conference on Genetic and Evolutionary Computation. ACM (2010)

How a Model Based on P-temporal Petri Nets Can Be Used to Study Aggregation Behavior

Fatima Debbat[1(✉)], Nicolas Monmarché[2], Pierre Gaucher[2],
and Mohamed Slimane[2]

[1] Computer Science Department, Mascara University Algeria, Mascara, Algeria
debbatfatima@gmail.com
[2] Laboratoire d'Informatique, University François Rabelais, Tours, France
{nicolas.monmarche,pierre.gaucher,mohamed.slimane}@univ-tours.fr

Abstract. In animal societies, many observed collective behaviours result from self-organized processes based on local interactions among individuals. Aggregation is widespread in insect societies and can appear in response to environmental heterogeneities or by attraction between individuals. Understanding this process requires linking individual behavioural rules of insects to a choice dynamics at the colony level. In this paper, we propose a model for the self-organized aggregation inspired by Jeason et al. aggregation behaviour model. Specifically, we use a probabilistic P-temporal Petri Nets model and analyse its performance using simulation. The results showed that this aggregation process, based on a small set of simple behavioural rules and interaction among individuals, can be used by the group of agent to select collectively an aggregation site among two identical or different shelters by estimating the size of each shelter during the collective decision-making process.

Keywords: Self-organization · Aggregation · Collective decision · Blattella germanica · Behavior model

1 Introduction

Since the last years, Swarm intelligence has undergone a considerable development. It refers generally to the study of the collective behavior of multi-component systems that coordinate using decentralized controls and self-organization. The self-organized processes are used by insect's society to make collective decisions: for instance in foraging in bees [1], of a nest site in ants [2], or of a shelter site in cockroaches [3]. The aggregation is one of these mechanisms; it results from interactions between individuals that follow simple rules based on local information, without reference to the global pattern. Aggregation is a step toward much more complex collective behaviours because it favours interactions and information exchanges among insects, leading to the emergence of complex and functional self-organized structures. As such it plays a key role in the evolution of cooperation in animal societies [4].

© Springer International Publishing Switzerland 2016
S. Bonnevay et al. (Eds.): EA 2015, LNCS 9554, pp. 191–204, 2016.
DOI: 10.1007/978-3-319-31471-6_15

Studies about aggregation can be grouped in three different but related fields; namely social insect studies, control theory and swarm robotics [5].

Ando *et al.* [6] introduced a deterministic algorithm for achieving aggregation in a group of mobile agents with limited perception in homogeneous environments. Cortés *et al.* [7] adapted this algorithm and showed that it can be used to achieve aggregation in arbitrarily high dimensions.

Deneubourg *et al.* [4] studied aggregation behavior in ants and cockroaches to form a bridge and cross a gap and cockroaches aggregate together in hiding sites. In these species, individuals rest in aggregations for varying time spans. The amount of time individuals spend in aggregations are modulated by environmental conditions and presence of other individuals. Individuals tend to spend more time in large aggregations, providing positive feedback for growth of aggregations. Individuals also spend more time on favorable sites, causing larger aggregates to form on such sites.

In one of the pioneering studies, Jeanson *et al.* [8] investigated aggregation in cockroach larvae, and developed a model of their behavior. The cockroaches were reported to join and leave clusters with probabilities correlated to the sizes of the clusters. They have demonstrated that the aggregation behaviour displayed by the German cockroach relies on a self-organization process: for a given moving cockroach, the larger the number of staying neighbors, the more likely the animal is to stop and stay beside them. This leads the cockroaches to quickly aggregate in dense clusters in a homogeneous environment. If one puts a dark shelter in a bright arena, one will observe that cockroaches strongly aggregate under this shelter. If two or more dark shelters are placed in the arena, a majority of the cockroaches will aggregate under a single shelter rather than evenly spreading among all resting sites. Garnier *et al.* [9] implemented this model with 20 Alice robots to achieve aggregation in homogeneous environments.

Amé *et al.* [10] set the probability of leaving a shelter based on a simple formula which makes the probability inversely proportional with the number of neighbors in that shelter. They give a system of differential equations describing the dynamical choice of a shelter in cockroaches. This system yields different qualitative collective behaviors, depending on the number and size of available shelters. Monte Carlo simulations showed that the formula is adequate for the individuals to select one of two shelters. In a subsequent study [11], Amé *et al.* developed a detailed model that took more than two shelters and the capacity of shelters into account.

Trianni [12] argues that "aggregation is of particular interest since it stands as a prerequisite for other forms of cooperation." In some cases, self-organized aggregation is aided by environmental heterogeneities, such as areas providing shelter or thermal energy.

Correll and Martinoli [13] analysed a similar model and showed that robots need a minimum combination of communication range and locomotion speed in order to aggregate into a single cluster when using probabilistic aggregation rules. They developed a probabilistic macroscopic model for aggregation. The model kept track of the number of robots in aggregates of specific size based on the waiting and leaving probabilities of robots and the encountering probability of searching robots with aggregates. The authors compared the results of a simulation model with the results of the macroscopic model and showed that the model predicts the aggregation dynamics successfully.

Soysal and Sahin [5] extended the controller in [8] by adding an approach behavior as a sub step prior to the wait behavior in which they can control the distance of a robot to its aggregate while waiting. A simulated robot which has infrared sensors for obtaining local information and a sound sensor/emitter pair for obtaining information from longer distances are used. Despite using a powerful sensing capability, the aggregations obtained were shown to be unscalable when the number of robots is increased.

This study supports evidence that aggregation relies on mechanisms of amplification, supported by interactions between individuals that follow simple rules based on local information and without knowledge of the global structure. We address self-organized aggregation behaviour of agents themselves based on P-temporal Petri Nets and inspired by Jeanson et al. model [8]. In this work, we investigate the potential of temporal Petri nets as a design/verification tool for self-organised systems. We show that a self-enhanced aggregation process, which leads groups of agents to a quick and strong aggregation, can be used by a group of agents to select collectively an aggregation site among two identical or different shelters.

We first describe the P-temporal Petri Nets model of aggregation we have used. We then develop an agent-based model implementing individual behavioural rules, to explain the aggregation dynamics at the collective level.

We show that, when this aggregation behaviour is restricted to certain zones in the environment, the agents preferentially aggregate in only one of these zones, i.e. they collectively choose a single shelter. When these shelters are of different sizes, the agents preferentially choose the biggest of the two, but without being individually able to measure their size.

2   Developed Behavioral Aggregation Model

To modulate the behavioral aggregation process, most of the researches have used probabilistic models or finite automaton state in which each of the discrete nodes corresponds to a distinct behavior. In order to model the aggregation behavior of the autonomous agents group, we have been inspired by the biological aggregation model developed by Jeason et al. [8] from experiments on the lips of the *Germanica* cockroach. As an improvement over the existing literature, in this work, we propose a P-temporal Petri Net model for managing the different states allowed for aggregation process. The Petri Net Theory (PNT) is well suited to describe the dynamic behaviour of complex concurrent systems based on graph theoretical concepts. It is also possible according to PNT to analyse properties of such systems. P-Temporal Petri nets are expected to be suitable for describing causal and temporal relationships between events of the aggregation behavior, including eventuality and fairness. By using Petri nets, the logical behavior of aggregation can be described by explicitly defining the causal relationships between events and probabilistic conditions in time.

A temporal Petri net (TPN) is presented by: $N = (P; T; FR; Eft; Lft; m0)$, Where:

- $P = P_1 \dots P_n$ is a finite set of places,
- $T = T_1 \dots T_m$ is a finite set of transitions,
- $FR \subseteq (P \times t) \cup (t \times P)$ the flow relation, ($\times$ operator is the cartesian product between places and transitions),
- Eft and Lft are the *earliest* and the *latest firing time* of the transitions respectively; $Eft(t) \leq Lft(t)$,
- $m0 \subseteq P$ is the initial marking.

This theory is modular, extensible and appropriate to demonstrate explicitly causal dependencies and independencies in a set of events and to illustrate the behaviour of a system on different levels of abstraction without changing tools and methods [15,16].

2.1   Aggregation Environment

The simulated environment is presented by a circular lighted arena containing shelters with variable size (dark places) and periphery zone. The agents are dispersed in arena and are considered to be identical. Each agent explores the arena randomly and his local environment is defined by circular perception area (blue circle in Fig. 1).

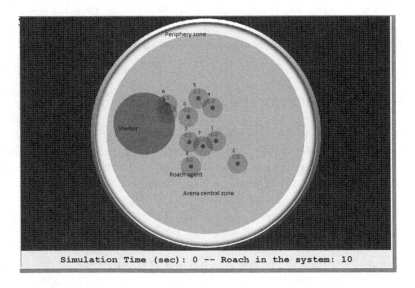

Simulation Time (sec): 0 -- Roach in the system: 10

Fig. 1. Aggregation Environment presentation (Color figure online)

2.2   Model Description

The total functionality of the group is decomposed into functional behaviors such as move or stop. The agents can be in the arena center, in the periphery or in the shelter. The transition probabilities for an agent to switch from one state to another are continuously modulated by its local environment within its perception radius, namely the number of stopped neighbours.

The proposed P-temporal Petri net (cf. Fig. 2) comprises 9 places and 20 transitions as given in Tables 1 and 2 respectively.

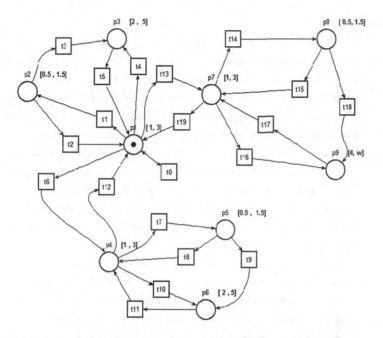

Fig. 2. P-temporal Petri nets Model

In arena centre, the agents start by random move state wandering aimlessly in the environment (P_1 place). They remain in this state by changing their direction at each period T (t_0 transition), if one of them detects one or more agents during his walk, it switches to the short stop state (P_2 place) with probability P_{short} and delay T_{short} or long stop state (P_3 place) with probability $(1 - P_{short})$ and delay T_{long}. This transition is conditioned by number of detected agents (one or more neighbors). If the stopped agent no longer detects other agents around itself, it switches back to move walk state (with a certain leave probability).

If an agent detects a periphery during his walk, it switches into a wall-following behavior (P_4 place) with probability $P_{periphery}$. In this state, the agent aligns its body with periphery and move and similarly at the arena centre, if it detects one or more agents, it switches into short stop state (P_5 place) with probability P_{short} and delay T_{short} or long stop state (P_6 place) with probability $(1 - P_{\text{short}})$ and delay T_{long}.

If an agent detects a shelter during his walk, it switches into an exploring shelter behavior (random walk P_7 place) with probability P_{shelter}. If it detects one or more agents, it switches into short stop state (P_8 place) with probability P_{short} and delay T_{short} or long stop state (P_9 place) with probability $(1 - P_{\text{short}})$ and great delay w.

Table 1. Description of transitions

Transition	Designation
T_0	Reflexive transition into move state with direction change
T_1	Transition into short stop state with probability P_{short} in arena centre
T_2	Transition into move state after $T_{\text{short}} \in [0.5, 1.5]$ in arena centre
T_3	Transition into long stop state after $T_{\text{short}} \in [0.5, 1.5]$ in arena centre
T_4	Transition into long stop state with probability $1 - P_{\text{short}}$ in arena centre
T_5	Transition into move state after $T_{\text{long}} \in [2, 5]$ in arena centre
T_6	Transition into move state in periphery with probability $P_{\text{periphery}}$
T_7	Transition into short stop state with probability P_{short} in periphery
T_8	Transition into move state after $T_{\text{short}} \in [0.5, 1.5]$ in periphery
T_9	Transition into long stop state after $T_{\text{short}} \in [0.5, 1.5]$ in periphery
T_{10}	Transition into long stop state with probability $1 - P_{\text{short}}$ in periphery
T_{11}	Transition into move state after $T_{\text{long}} \in [2, 5]$ in periphery
T_{12}	Transition into move state in arena centre with probability $P_{\text{periphery}}$
T_{13}	Transition into move state in shelter with probability P_{shelter}
T_{14}	Transition into short stop state with probability P_{short} in shelter
T_{15}	Transition into move state after $T_{\text{short}} \in [0.5, 1.5]$ in shelter
T_{16}	Transition into long stop state with probability $1 - P_{\text{short}}$ in shelter
T_{17}	Transition into move state after great $T_{\text{long}} \in [4, w]$ in shelter
T_{18}	Transition into long stop state after $T_{\text{short}} \in [0.5, 1.5]$ in shelter
T_{19}	Transition into move state in arena centre with probability $(1 - P_{\text{shelter}})$

The values of the empirical transition probabilities are $P_{\text{short}} = 0.1$, $P_{\text{periphery}} = 0.3$ and $P_{\text{shelter}} = 0.7$.

The proposed Petri Net model should be bounded and live in time (Liveness and Boundness: are the dynamic properties of each Petri Net). Before testing these properties, we give some definitions [14]:

- **Boundless and Safeness**: A Petri-Net is k-bounded if for every reachable marking the number of tokens in any place is not greater than k (a place is called k-bounded if for every reachable marking the number of tokens in it is

Table 2. Description of places

Places	Designation
P_1	Move state in arena centre with delay $T \in [1,3]$
P_2	Short stop state in arena centre with delay $T_{short} \in [0.5, 1.5]$
P_3	Short stop state in arena centre with delay $T_{long} \in [2,5]$
P_4	Move state in periphery with delay $T \in [1,3]$
P_5	Short stop state periphery with delay $T_{short} \in [0.5, 1.5]$
P_6	Short stop state periphery with delay $T_{long} \in [2,5]$
P_7	Move state in shelter with delay $T \in [1,3]$
P_8	Short stop state shelter with delay $T_{short} \in [0.5, 1.5]$
P_9	Short stop state shelter with delay $T_{long} \in [4,w]$

not greater than k). A Petri-Net is bounded, if there is a finite k for which it is k-bounded. A Petri-Net is safe if it is 1-bounded (1 bounded place is called a safe place).
- **Liveness**: A Petri-Net is live if for every transition t and every reachable marking M there is a firing sequence that leads to a marking M' enabling t.

In order to check these properties, we have applied the simulation method by using TINA. TINA (Time petri Net Analyser) is a software environment to edit and analyse Petri nets, Time Petri nets, and some extensions of these nets [16]. We can get with the help of TINA the verification of correctness of the proposed design. Figure 3 presents the analysis results of the proposed Petri Net. Analysis with TINA tool confirms the liveness, reachability of system states, invariants, and no deadlocks of the proposed Petri Net system. The petri Network is bounded and all transitions are live from a given initial marking and agent configuration in environment. It shows that the agent will not be in a deadlock or infinite cycle during its aggregation.

3   Simulation and Results

In this section, we developed a simulator to study the relevance of the proposed model based on P-temporal Petri Nets. We use Java language with the standard API J2D and JFreeChart library. The objective of simulations is to bring the agents together at some location in the environment (i.e. aggregate them) via P-temporal Petri Nets control. Each agent is equipped with a sensor, which allows it to know whether or not there is another agent in the direct line of sight of the sensor. Agents have an average speed of 1 cm/s and can evaluate the number of neighbors using local communication (we assume that each robot has unique ID or can randomly choose its ID from a sufficiently large set). Notice that behavioural probabilities in our model are independent from the agents speed. The communication range (here 1 cm) of each individual is depicted by a circular blue disc. The number of agents is changed in the simulations in order to analyze the scalability of the approach.

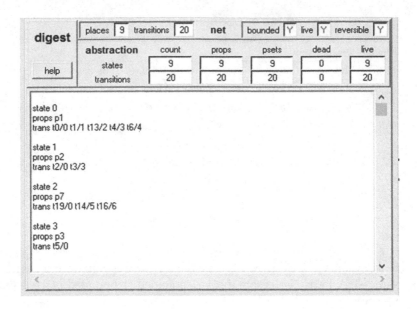

Fig. 3. Reachability analysis of P-temporal Petri-Nets model

The aggregation dynamics were characterized through three kinds of metrics: size of shelter, number of Shelters (without, one or two shelter) and number of isolated agents in arena. In the first case, we achieved the implementation of our model in environment without shelter (Fig. 4). 12 agents are placed in

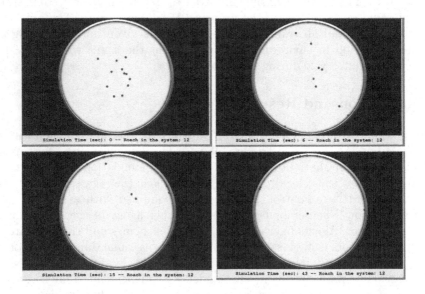

Fig. 4. Snapshots of simulation for aggregation process in arena periphery

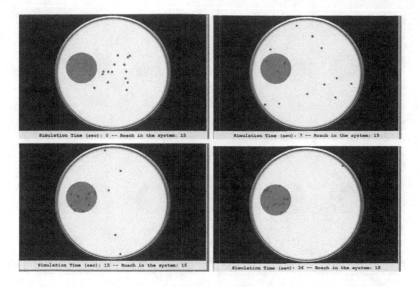

Fig. 5. Snapshots of simulation for aggregation process in Shelter

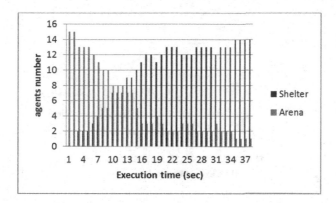

Fig. 6. Choice dynamics: number of agents aggregated under shelter

a homogeneous arena, with random positions and orientations. After 40 s, we observe that agents strongly aggregate under the periphery (one isolated agent).

In the second case, we put one shelter in arena (dark place) with 15 agents (Fig. 5). We can observe that agents prefer and aggregate under this shelter (after 36 s, 14 agents are under shelter and only one in periphery).

In this simulation, the number of agents in shelter was counted every one second to characterize the aggregation dynamics under each shelter (Fig. 6).

In Fig. 7, the snapshots correspond to the simulation with two identical shelters and 15 agents. Note that a agent can be in one of these three locations: under shelter 1, under shelter 2 or outside the shelters. As can be seen, the simulation ended with the choice of one of the two shelters by agents.

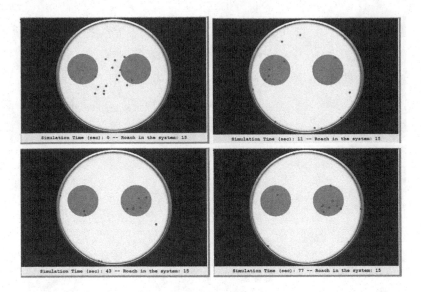

Fig. 7. Snapshots of simulation for aggregation process in two identical Shelters

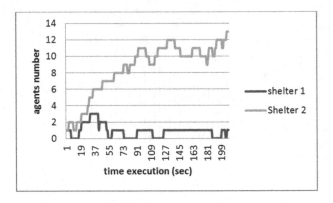

Fig. 8. Choice dynamics: number of agents aggregated under two identical Shelters

The agents are able to perform a collective choice for a given aggregation shelter, even if these shelters are identical (Fig. 8).

In the following simulation case (Fig. 9), we put two different shelters in size (shelter 1 diameter is 2.5 cm and shelter 2 diameter is 4.5 cm) and 15 agents.

In this case, we showed that a self-enhanced aggregation process is associated with a preference for a largest shelter. The agents strongly choose the shelter able to house their whole population. Furthermore, this choice can be related to a collective ability to compare the sizes of the aggregation sites.

In the last case of simulation (Fig. 10), we put two identical smallest shelters (diameter for each shelter is 2 cm) with 20 agents.

As can be seen, the simulation ended with the choice for the two shelters by agents (with equivalent partition).

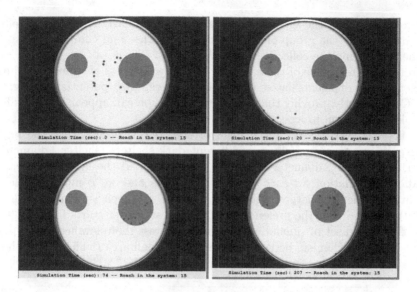

Fig. 9. Snapshots of simulation for aggregation process in two different Shelters

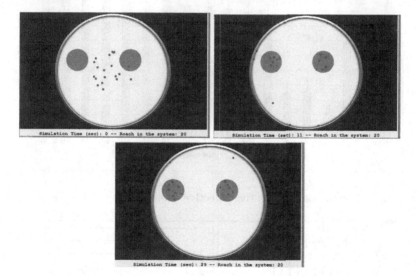

Fig. 10. Snapshots of simulation for aggregation process in two identical smallest Shelters

From these sets of simulation, we can conclude that the group of agent will choose preferentially a shelter that is sufficiently large to house all its members. But when the group is confronted with two sufficiently large shelters, the self-enhanced aggregation mechanism can lead the group to two stable choices, with a preference for the larger shelter. If the two shelters are not able to house the whole of population, the agents can be partitioned similarly in the both shelters. This implies that the group of robots is able to sense and compare the size of the shelters during the collective decision process.

To test the relevance of these results, we ran a two sets of simulation during which a group of agents faced the choice between two potential aggregation sites. Consequently, proving that a collective decision can appear in agents from a simple aggregation process.

In first set, we test the mechanism of the collective choice for the largest shelter. We use an arena with two different dark shelters (shelter 1 diameter is 2.5 cm and shelter 2 diameter is 4.5 cm) and we varied the agent number. We repeat the simulation for each number of agents and we compute the agent percentage under the largest shelter at the end for each simulation (Figs. 11 and 12). We find that the preference of the largest shelter starts from 9 agents.

In the second set of simulation (30 runs), we test the relevance of the population sharing mechanism between shelters with small area (which cannot house

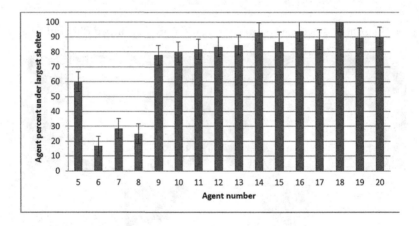

Fig. 11. Simulation set for the collective choice for the largest shelter

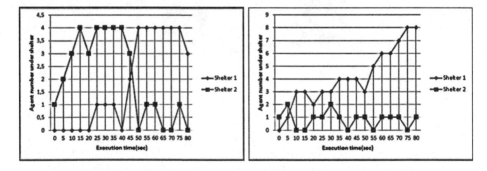

Fig. 12. Choice dynamics: aggregation under two different shelters with varied agent number (5 in row 1 and 15 in row 2)

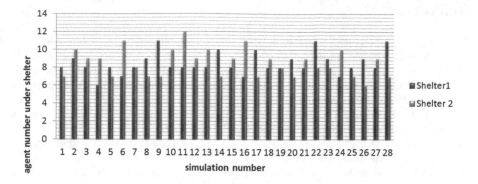

Fig. 13. Simulation set for the population sharing mechanism between shelters

the whole population of agents). We use an arena with two identical dark shelters (shelter 1 diameter is 2.5 cm and shelter 2 diameter is 2.5 cm) and 20 agents. We compute the agent number in each shelter at the end of simulation for each run (Fig. 13). For the whole execution, we notice an equivalent sharing between the two shelters. The average agent in the shelter 1 is 8.46 and for the shelter 2 is equal to 8.5, which confirms this mechanism of self-enhanced in aggregation process.

4   Conclusion

In this work, we achieved development and implementation of a behavior aggregation model based on P-temporal Petri Nets in a group of agents that is capable of quantitative prediction of self-organized agent aggregation dynamics. This model is tested with sets of simulation in different environment configurations (without shelter, one shelter, two identical shelters with different size and two different shelters). In terms of sensory abilities between biological and the proposed model, the aggregation dynamics observed in agents closely match those observed in cockroaches (Jeason et al. model). This proposed model attested that, the choice of aggregating site can be related to a collective ability to compare the sizes of the aggregation sites. This work opens some interesting perspectives. We project to extend this model for self-organized phenomena based on P-temporel Petri Nets on other scales that might be governed by similar or different behaviors of swarm intelligence by using of evolutionary approaches.

References

1. Seley, T., Camazine, S., Sneyd, J.: Collective decision-making in honey bees: how colonies choose among nectar sources. Behav. Ecol. Sociobiol. **28**, 277–290 (1991)
2. Jeanson, R., Deneubourg, J.-L., Theraulaz, G.: Modulation of individual behavior and collective decision-making during aggregation site selection by the ant Messor sancta. Behav. Ecol. Sociobiol. **55**, 388–394 (2004)
3. Amé, J.-M., Rivault, C., Deneubourg, J.-L.: Cockroach aggregation based on strain odour recognition. Anim. Behav. **68**(4), 793–801 (2004)
4. Deneubourg, J.L., Lioni, A., Detrain, C.: Dynamics of aggregation and emergence of cooperation. Biol. Bull. **202**(3), 262–267 (2002)
5. Soysal, O., Şahin, E.: A macroscopic model for self-organized aggregation in swarm robotic systems. In: Şahin, E., Spears, W.M., Winfield, A.F.T. (eds.) SAB 2006 Ws 2007. LNCS, vol. 4433, pp. 27–42. Springer, Heidelberg (2007)
6. Ando, H., Oasa, Y., Suzuki, I., Yamashita, M.: Distributed memoryless point convergence algorithm for mobile robots with limited visibility. IEEE Trans. Rob. Autom. **15**(5), 818–828 (1999)
7. Cortés, J., Martinez, S., Bullo, F.: Robust rendezvous for mobile autonomous agents via proximity graphs in arbitrary dimensions. IEEE Trans. Autom. Control **51**(8), 1289–1298 (2006)
8. Jeanson, R., Rivault, C., Deneubourg, J.-L., Blancos, S., Fourniers, R., Jost, C., Theraulaz, G.: Self-organized aggregation in cockroaches. Anim. Behav. **69**, 169–180 (2005)
9. Garnier, S., Jost, C., Gautrais, J., Asadpour, M., Caprari, G., Jeanson, R., Grimal, A., Theraulaz, G.: The embodiment of cockroach aggregation behavior in a group of micro-robots. Artif. Life **14**(4), 387–408 (2008)
10. Amé, J.-M., Rivault, C., Deneubourg, J.-L.: Cockroach aggregation based on strain odour recognition. Anim. Behav. **68**, 793–801 (2004)
11. Amé, J.-M., Halloy, J., Rivault, C., Detrain, C., Deneubourg, J.-L.: Collegial decision making based on social amplification leads to optimal group formation. Proc. Nat. Acad. Sci. USA **103**, 5835–5840 (2006)
12. Trianni, V.: Evolutionary Swarm Robotics. Studies in Computational Intelligence, vol. 108. Springer, Berlin (2008)
13. Correll, N., Martinoli, A.: Modeling and designing self-organized aggregation in a swarm of miniature robots. Int. J. Robot. Res. **30**(5), 615–626 (2011)
14. Herrmann, J.W., Lin, E.: Tutorial and Applications. In: The 32th Annual Symposium of the Washington Operations Research Management Science Council (1997)
15. Berthomieu B., Vernadat F.: Time petri nets analysis with TINA, tool paper, In: Proceedings of 3rd International Conference on The Quantitative Evaluation of Systems (QEST 2006). IEEE Computer Society (2006)
16. TINA toolbox. http://projects.laas.fr/tina/. (access on 11 November 2015)

A Distributed Hybrid Algorithm for the Graph Coloring Problem

Ines Sghir[1,3], Jin-Kao Hao[1,2(✉)], Ines Ben Jaafar[3], and Khaled Ghédira[3]

[1] LERIA, Université d'Angers, 2 Bd Lavoisier, 49045 Angers Cedex 01, France
jin-kao.hao@univ-angers.fr
[2] Institut Universitaire de France, Paris, France
[3] SOIE, ISG, Université de Tunis, Cité Bouchoucha 2000 Le Bardo, Tunis, Tunisia

Abstract. We propose a multi-agent based Distributed Hybrid algorithm for the Graph Coloring Problem (DH-GCP). DH-GCP applies a tabu search procedure with two different neighborhood structures for its intensification. To diversify the search into unexplored promising regions, two crossover operators and two types of perturbation moves are performed. All these search components are managed by a multi-agent model which uses reinforcement learning for decision making. The performance of the proposed algorithm is evaluated on well-known DIMACS benchmark instances.

1 Introduction

Given an undirected graph $G = (V, E)$ with vertex set V and edge set E. A legal (or proper) k-coloring of G (k is an integer) is a partition of V, i.e., $S = \{V_1, V_2, ..., V_k\}$ where each subset $V_r \subset V$ is an independent set (also called a legal color class) such that no two vertices of V_r are linked by an edge. Given k colors, the k-coloring problem (k-COL) is to find a legal k-coloring. The graph coloring problem (GCP) is to determine the smallest integer k (i.e., the chromatic number χ_G of G) such that there exists a legal k-coloring of G.

GCP has numerous important applications in practice and is known to be computational difficult. Given its relevance, GCP is certainly among the most studied NP-hard problems [8]. Among the large number of GCP solution approaches (see e.g., [12,20]), most of them are based on neighborhood search [1,2,5,6,15–17,22,27], hybrid population search [4,7,9,10,18,19,21,23,26] or other hybrid scheme [14,24,28]. More GCP methods can be found in [8,12,20].

In this paper, we study a distributed algorithm for GCP which is based on the principle of multi-agent systems. As our general solution strategy, we adopt the very popular k-fixed penalty approach [12] which was used in many previous algorithms like [5,6,9,12,18,19,23]. With this approach, we fix the number k of colors and seek a legal k-coloring among all possible (legal or illegal) k-colorings. Given a k-coloring S, the evaluation or fitness function $f(S)$ calculates the number of conflicts induced by S, i.e., the number of edges whose end-points are colored with the same color. Thus, $f(S) = 0$ indicates that S is a legal coloring.

© Springer International Publishing Switzerland 2016
S. Bonnevay et al. (Eds.): EA 2015, LNCS 9554, pp. 205–218, 2016.
DOI: 10.1007/978-3-319-31471-6_16

The algorithm tries to solve the k-coloring problem by minimizing the fitness function f. Finally, to approximate the chromatic number of G, we try to solve a series of k-coloring problems with decreasing values of k.

The rest of the paper is organized as follows. Section 2 describes the proposed algorithm. Section 3 presents the experimental results achieved on DIMACS benchmark instances. Finally, Sect. 4 concludes the paper.

2   A Distributed Hybrid Algorithm for GCP

The proposed distributed hybrid algorithm for GCP (DH-GCP) explores a set of interacting agents which are local optimization procedures, crossover operators and perturbation techniques. The coordination of these agents is realized in an informed way using reinforcement learning. The learning mechanism modifies and adapts the search strategy according to the experiences obtained during the search process. The agents are learners and players that ensure the role of intensification and diversification to explore the given search space. This study constitutes a continuation of our recent work on multi-agent based optimization applied to the quadratic assignment problem [25].

2.1   Weight Matrix with Reinforcement Learning

Reinforcement learning aims to learn what to do and how to plan situations to actions, in order to maximize a numerical reward signal. In most forms of learning, the learner is told which actions to take, but for reinforcement learning, the learner needs to discover the action that leads to the best reward based on previous experiences. A learner must be able to learn from its own experiences to make decisions. In the proposed DH-GCP algorithm, decisions or actions correspond to techniques of diversification or intensification to apply and experiences are acquired during the search progress. Following [13], we use decision rules represented by a couple $(Condition, Action)$. Let C be the set of conditions describing the search progress and A the set of actions or decisions to perform. For a condition C_i, a weight W_{ij} (initialized to 0) is associated to each action A_j. We use the following equation [13] to calculate the probability $P(C_i, A_j)$ for applying an action A_j based on a condition C_i:

$$P(C_i, A_j) = \frac{W_{ij}}{\sum_{j \in A} W_{ij}} \tag{1}$$

At the beginning of the algorithm (i.e., first iterations of Algorithm 2), the improvement situation is assigned to a default condition. According to the weight matrix W, the most appropriate action for this condition is selected based on the probability given in Eq. (1). Then, at the end of each generation, the performed action is evaluated and the concerned weight value is increased if there is an

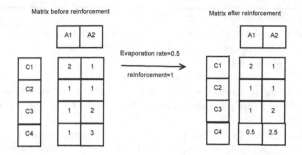

Fig. 1. An example of the reinforcement learning procedure with weight matrix: We suppose that the current condition is C_4 (e.g., the local best solution has not been improved in recent 10 generations). Under this condition, action A_2 (e.g., activate crossover agents) is performed for the current generation (this action has the highest value in the matrix) and obtained a further improvement. Then, reinforcement is applied by adjusting the weight W_{42} to augment the chance of selecting again the applied action under this condition (e.g. $W_{42} = 3 \times 0.5 + 1 = 2.5$). The weight W_{41} is decreased by μ (e.g. $W_{41} = 1 \times 0.5 = 0.5$)

improvement in solution quality. A credit assignment is used to perform reinforcement learning in order to select the beneficial experiences and determine a reward for them. Here, an experience is represented as a triplet (*condition C_i*, *action A_j*, *improvement V*). When a new best local or global solution is found, the weight value W_{ij} which is related to the action of this generation is reinforced by adding a reward rate σ to W_{ij}. Before adding the reinforcement value, the weight values W_{ij} in the decision matrix is decreased with an evaporation value μ, in order to enlarge the influence of the new reward obtained in the current generation. The reinforcement with reward σ is then performed using the following equations [13]:

$$W'_{ij} = \mu \times W''_{ij} \tag{2}$$

$$W_{ij} = \mu \times W'_{ij} + \sigma \tag{3}$$

where W'_{ij} is the weight value before the reinforcement, W''_{ij} is the weight value before the evaporation, μ is the evaporation value and σ is the learning factor.

Figure 1 shows an illustrative example of this reinforcement learning process (See Sect. 2.3 for more details). In the proposed algorithm, this matrix is used by the mediator agent (Sect. 2.3) and the tabu search agents (Sect. 2.4).

2.2   Agent Interaction in DH-GCP

The proposed DH-GCP is a distributed approach composed of interacting agents. Each agent has a local view of the problem, but the collaboration of

---

Algorithm 1. DH-GCP general procedure

Require: Graph G, number of colors k, four types of agents: mediator agent, two tabu search agents, perturbation agent and two crossover agents
Ensure: The best k-coloring S_{best}
 1: **while** A legal k-coloring S_{best} not reached **do**
 2:    The mediator agent starts the algorithm by initialing the search and then decides to trigger tabu search agents or crossover agents based on its weight matrix (Algorithm 2)
 3:    **if** The mediator agent decides to activate tabu search agents **then**
 4:        The tabu search agents are activated and the mediator agent waits for a k-coloring from the activated tabu agents (Algorithm 3)
 5:        **if** An activated tabu search agent needs to trigger a perturbation agent **then**
 6:            The tabu search agent actives the required perturbation agent and waits for solution from the perturbation agent (Sect. 2.5)
 7:            The perturbation agent is killed after sending the k-coloring found to the corresponding tabu search agent
 8:        **end if**
 9:        **if** An tabu search agent wants to cooperate with other tabu search agent **then**
10:            The requiring tabu search agent waits for a new k-coloring from other tabu search agent (Algorithm 3)
11:        **end if**
12:        The tabu search agents are killed after sending the best solutions generated during their search to the mediator agent
13:    **end if**
14:    **if** The mediator agent decides to activate crossover agents **then**
15:        The crossover agents are activated and the mediator agent waits for the best k-coloring from the crossover agents (Sect. 2.6)
16:        The crossover agents are killed after sending new solutions to the mediator agent
17:    **end if**
18: **end while**
19: **Return** The best legal k-coloring found S_{best} from the mediator agent

---

these agents can help find good solutions for GCP. We consider the following agents: the mediator agent, the tabu search agents, the perturbation agent and the crossover agents. Figure 2 describes the architecture of DH-GCP while Algorithm 1 presents the general procedure of DH-GCP. Algorithms 2 and 3 describe the behaviors of the mediator agent and the tabu search agents.

The DH-GCP algorithm explores several search cycles (generations, see the 'while' structure of Algorithm 2). In each cycle, the mediator agent is responsible to decide which agents will be activated using its weight matrix according to the state of search process. The activated agents can be tabu search agents or crossover agents (Algorithm 2). During the process of tabu search agents, they can trigger the perturbation agent (to diversify the search). Note that agents are not activated in a pre-specified order. Instead, their activation depends

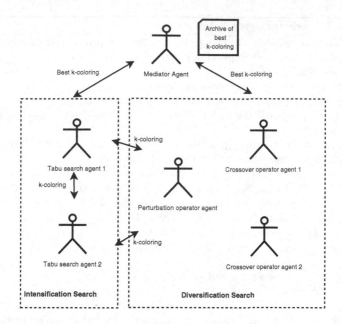

Fig. 2. Agent communication in DH-GCP: Mediator agent is the agent who manages the search according to the improvement realized, tabu search agents ensure search intensification while crossover agents and perturbation agent are responsible for diversification

on the past learning experiences and is dynamically adjusted. In the following subsections, we explain the behaviors of each type of agents.

2.3   Mediator Agent

The mediator agent selects other agents to trigger based on its weight matrix (Sect. 2.1). When other agents (tabu search agents or crossover agents) are triggered, the mediator agent waits for (improved) solutions received from these agents and to record the received solutions in the shared memory (archive). The behavior of the mediator agent is described in Algorithm 2.

The Initial Solution. The mediator agent creates an initial legal coloring using the greedy largest saturation degree heuristic (DSATUR) [3]. Then, starting with this initial coloring, it randomly displaces the vertices whose color number is higher than the given color number k to a color class between $[1, k]$. This procedure usually leads to an illegal k-coloring which will be repaired by the DH-GCP algorithm.

Conditions and Actions of Weight Matrix. The weight matrix of the mediator agent is composed of two types of actions: A_1 corresponds to activating

---

Algorithm 2. Mediator agent behavior

Require: Graph G, number of class k, parameters: improvement threshold interval *interval*, consecutive non-improving iterations *max_opt*.

Ensure: A best legal k-coloring S_{best} found so far

1: $S \leftarrow Generate\_initial\_k - coloring\_S_0$ {Sect. 2.3}
2: $S_{best} \leftarrow S_0$ {S_{best} records the best k-coloring found so far}
3: $f_{best} \leftarrow f_0$ {f_{best} records the best objective value of the best k-coloring reached so far}
4: $opt \leftarrow 0$ {opt is the counter for consecutive non-improving local optimum}
5: $W \leftarrow 0$ {W is the weight matrix of the mediator agent}
6: $pop \leftarrow \emptyset$ {pop is the archive of elite solutions found during the search}
7: **while** A legal k-coloring S_{best} not reached ($f_{best} \neq 0$) **do**
8:     Update W using *interval*, opt and *max_opt* {Sects. 2.1 and 2.3}
9:     $Action\_type \leftarrow$ Select an action to activate based on W {Sect. 2.3}
10:    **if** $Action\_type =$ tabu search agents **then**
11:       Trigger tabu search agents and send S_{best} to them
12:    **else**
13:       Trigger crossover agents and send S_{best} to them
14:       $opt \leftarrow 0$
15:    **end if**
16:    $S_1 \leftarrow \emptyset$, $S_2 \leftarrow \emptyset$ {S_1 and S_2 are k-colorings received from the activated agents}
17:    **if** $S_1 \neq \emptyset$ AND $S_2 \neq \emptyset$ **then**
18:       **if** $f(S_1) \leq f(S_2)$ **then**
19:         $S \leftarrow S_1$
20:       **else**
21:         $S \leftarrow S_2$
22:       **end if**
23:       $tr \leftarrow Exist(S_1, S_2, pop)$ {Check if S_1 and/or S_2 are in the archive pop}
24:       **if** $tr =$ false **then**
25:         add S_1 and/or S_2 to pop {add both k-colorings or one of them in pop}
26:       **end if**
27:       **if** $f(S) \leq f_{best}$ **then**
28:         $S_{best} \leftarrow S$
29:         $f_{best} \leftarrow f(S)$
30:       **else**
31:         $opt \leftarrow opt + 1$
32:       **end if**
33:    **else**
34:       block this agent {The mediator agent waits for k-coloring from other activated agents}
35:    **end if**
36: **end while**
37: **Return** S_{best}

---

the tabu search agents, and A_2 corresponds to activating the crossover agents. The conditions, which cover significant situations that may occur in the search process, are:

- C_1 = The algorithm does not reach m_0 generations (cycles);
- C_2 = The local or global best solution is improved in recent m_1 generations and this improvement is a small improvement in the fitness function value f;
- C_3 = The local or global best solution is improved in recent m_1 generations and this improvement is a large improvement in the fitness function value f;
- C_4 = The global best solution has not been improved in recent m_2 generations. This solution is a deep local optimum or an optimum solution.

where m_0, m_1 and m_2 are parameters set by the user according to the total generation number. When there is a large improvement obtained by the application of an action between two successive generations (this corresponds to the situations C_1 and C_3), it is better to apply an intensification process by triggering the tabu search agents. If the mediator agent observes no improvement or an insignificant improvement (this corresponds to the situations C_2 and C_4), the search needs to be diversified by activating crossover agents. After each generation (i.e., when the activated agents return their found solution), the mediator agent updates its weight matrix (see Sect. 2.1).

Archive of Elite Solutions. The mediator agent saves the best k-coloring, received from tabu search agents and crossover agents, in an archive. The archive represents a shared memory between all agents. It is updated by the mediator agent with new solutions of good quality.

2.4   Tabu Search Agents

The mediator agent can activate two tabu search agents, when it observes that the search process needs to be intensified (lines $4 - 11$ of Algorithm 2). Each tabu search agent applies a specific strategy based on a particular neighborhood to seek new solutions (line 7 to line 10 of Algorithm 3). During the search, a tabu search agent can exchange its solutions with another alive tabu search agent or with a perturbation agent (line 14 to line 28 of Algorithm 3). These communications depend on a weight matrix(lines 16 and 17 of Algorithm 3). At the end of each tabu search agent run, the best k-coloring found by each agent is sent to the mediator agent (line 36 of Algorithm 3). The behavior of the tabu search agent is described in Algorithm 3. Below, we define the used neighborhood structures for each tabu search agent. Then, we explain the conditions and actions employed by them.

Neighborhoods. A candidate solution for GCP can be generated by changing the color class of vertices. Different modifications lead to different neighborhood structures. In this work, we explore 3 neighborhoods: the vertex neighborhood which changes the color of some conflicting vertices, the class neighborhood which changes the color of some or all vertices of a conflicting color class, and the non-increasing neighborhood which changes the color of some vertices without increasing the total number of conflicting edges.

Algorithm 3. Tabu search agents behavior

Require: Graph G, number of colors k, A k-coloring S_{best} received from mediator agent, parameters: maximum iterations $iteration\_max$, improvement threshold
interval $interval$, consecutive non-improving iterations max_opt_TS.
Ensure: A k-coloring S_{best_TS}

1: $S \leftarrow S_0$ {S is the current k-coloring found by each tabu search agent}
2: $Tabu\_list \leftarrow 0$ {$Tabu\_list$ is the tabu list, Sect. 2.4}
3: $Q \leftarrow 0$ {Q is the weight matrix of each tabu search agent, Sect. 2.4}
4: $opt = 0$ {opt is the counter for consecutive non-improving local optima for each
    tabu search agent}
5: $S_1 \leftarrow S_0$ {S_1 is the k-coloring obtained in generation $iteration - 1$}
6: **while** $iteration \leq iteration\_max$ **do**
7:     $S \leftarrow$ Generate the best neighboring k-coloring {Sect. 2.4}
8:     Update $Tabu\_list$
9:     **if** $f(S) \leq f(S_{best\_TS})$ **then**
10:         $S_{best\_TS} \leftarrow S$
11:     **else**
12:         $opt = opt + 1$
13:     **end if**
14:     **if** $(f(S) - f(S_1) < interval)$or $(opt = max\_opt\_TS)$ **then**
15:         $S_{perturbed} \leftarrow \emptyset$ {$S_{perturbed}$ is a k-coloring received from other agents (tabu
            search agent or perturbation agent}
16:         Update Q {Update the weight matrix based on the improvement of the current
            solution, opt, $interval$ and max_opt_TS, Sect. 2}
17:         $Action\_exchange \leftarrow$ Select the agent to trigger based on Q
18:         **if** $Action\_exchange$=Activating perturbation agent **then**
19:             Activate the perturbation agent and send it the current k-coloring S
20:             opt=0 {opt is reset to 0, only when strong perturbation behaviour is trigged,
                Sect. 2.5}
21:         **end if**
22:         **if** $Action\_exchange$=Activating other tabu search agent **then**
23:             Request the current k-coloring of other tabu search agent
24:         **end if**
25:         Let $S_{perturbed}$ be the perturbed solution received from the perturbation agent
            or other tabu search agent
26:         **if** $S_{perturbed} \neq \emptyset$ **then**
27:             $S \leftarrow S_{perturbed}$
28:         **else**
29:             block this agent {Tabu search agent waits for a solution from other agents}
30:         **end if**
31:     **else**
32:         $S_1 \leftarrow S$ {Tabu search agent applies tabu search without exchanging solutions
            with other agents}
33:     **end if**
34:     $iteration = iteration + 1$
35: **end while**
36: **Return** S_{best_TS} to mediator agent

Neighborhood Exploration Strategies. In DH-GCP, we use two comple-
mentary neighborhood strategies due to the cooperation act realized by each
tabu search agent. One of these strategies, performed by our first tabu search
agent, changes the colors of conflicting vertices to produce new k-colorings. This
is done by moving a conflicting vertex x from its original color class V_i to the best
possible other color class V_j $(i \neq j)$ (this change or move is denoted by (x, i, j)).
The new color class for each conflicting vertex x is chosen among those which are
not assigned to vertices adjacent to x. Among these color classes found, the best
possible color class (in terms of fitness minimization) is selected for the considered conflicting vertex. Our second tabu search age nt uses the same mechanism
of selecting the best color class to be assigned to vertices as the first tabu search
agent. The difference is that these vertices are not the set of conflicting vertices,
but the adjacent of conflicting vertices. The tabu search agent chooses the best
color class for each vertex belonging to the set of adjacent vertices of conflict
vertices. The best color affected must not belong to the color classes affected to
conflicting vertices.

For these two neighborhood strategies, tabu search agents evaluate each move
using an incremental evaluation technique [6,9,10]. This technique consists of
maintaining a special data structure that records the move values for each candidate neighborhood move.

Tabu List. Each tabu search agent uses a tabu list to forbid the reverse moves.
When a move (x,i,j) is generated, vertex x is forbidden to move back to color
class V_i for the next h iterations (called tabu tenure). The tabu tenure is dynamically determined by $h = f(S) + r(10)$ where $r(10)$ is a random number between
1 and 10 [10]. The stop condition of each tabu search is a fixed number of iterations.

Conditions and Actions of Weight Matrix. The actions considered by the
tabu search agents are as follows:

- A_1 = activating other tabu search agents;
- A_2 = activating the strong perturbation behavior in the perturbation agent;
- A_3 = activating the reduced perturbation behavior in the perturbation agent.

The set of the conditions are:

- C_1 = the local best k-coloring is improved in recent q_1 generations and this
  improvement is a small improvement in the fitness function value f;
- C_2 = the local best k-coloring is improved in recent q_2 generations;
- C_3 = the local best k-coloring is improved in recent q_3 generations and $q_3 > q_2$.

where q_1, q_2 and q_3 are parameters set by the user according to the total generation number.

Each of these conditions promotes a certain action. Thus, C_1 increases the chance of activating other tabu search agent, to reinforce intensification. C_2 and C_3 reinforce the action of triggering the perturbation agent, in order to increase diversification. The selection of the most suitable action is controlled by the corresponding weight matrix of each tabu search agent.

2.5   Perturbation Operator Agent

The perturbation agent, triggered by tabu search agents (lines 18–20 of Algorithm 3), creates a disturbed k-coloring solution by exploring two types of perturbations. The new k-coloring is then sent to the tabu search agent (line 25 of Algorithm 3) for further improvement.

Reduced Perturbation Technique. The reduced perturbation technique can be triggered when a tabu search agent observes a slight search stagnation (condition C_2 of Sect. 2.4). From the k-coloring received from the tabu search agent, the perturbation agent makes t moves to create a new solution, where each move changes randomly the color of a conflicting vertex of the incumbent solution. The number t of moves is chosen randomly between 1 and $conf$ (where $conf$ is the number of conflicting vertices).

Strong Perturbation Technique. The strong perturbation technique is performed when a tabu search agent tabu search agent observes deep search stagnation. The perturbation agent uses the shared archive of elite k-colorings to create a new solution. It extracts the number of occurrence of each vertex x colored by each color class V_i. Starting with an uncolored graph, each vertex x is colored with a color class V_i which has the smallest occurrence number. Dedicated data structures are employed to avoid the creation of the same solution for future calls to the perturbation agent.

2.6   Crossover Agents

When the mediator agent decides to activate the crossover agents (line 13 of Algorithm 2), two crossover agents are created based on two different operators from the literature: the AMPaX operator [18] and the GPX operator [10]. The new k-coloring solutions from the two crossover agents are sent to the mediator agent to continue the search process. Experimental results showed that the joint use of these two crossover operators performs better than any of them used alone.

3   Experimentation

3.1   Experimental Results

In this section, we present experimental results of our DH-GCP algorithm on a set of well-known DIMACS coloring benchmarks and compare the results with

Table 1. Computational results of DH-GCP on the set of difficult DIMACS challenge benchmarks

Instances	$|V|$	$|E|$	dens	k^*	References	DH-GCP		
						k	hit	time(m)
DSJC250.5	250	15,668	0.50	28	[10, 11, 14, 16, 19, 22, 23, 26, 28]	28	10/10	5
DSJC500.1	500	12,458	0.10	12	[2, 11, 14, 16, 19, 21–23, 26, 28]	12	10/10	6
DSJC500.5	500	62,624	0.50	47	[21]	-	-	-
				48	[2, 10, 11, 14, 16, 19, 22, 26, 28]	48	10/10	85
DSJC500.9	500	112,437	0.90	126	[2, 11, 14, 16, 19, 21–23, 26, 28]	126	10/10	320
DSJC1000.1	1000	49,629	0.10	20	[2, 10, 11, 14, 16, 19, 21–23, 26, 28]	20	10/10	441
DSJC1000.5	1000	249,826	0.5	82	[21]	-	-	-
				83	[10, 14, 19, 22, 23, 28]	83	10/10	205
DSJC1000.9	1000	449,449	0.90	222	[2, 10, 14, 16, 21–23, 26, 28]	222	4/10	801
DSJR500.1c	500	121,275	0.97	85	[14, 16, 19, 23, 26, 28]	85	10/1	6
DSJR500.5	500	58,862	0.47	122	[14, 16, 23, 24, 26, 28]	122	3/10	480
R250.5	250	14,849	0.48	65	[2, 14, 19, 23, 26, 28]	65	10/10	42
R1000.1c	1000	485,090	0.97	98	[2, 14, 19, 23, 26, 28]	98	10/10	55
R1000.5	1000	238,267	0.48	234	[16]	-	-	-
				238	[26]	240	2/10	1120
le450_15c	450	16,680	0.17	15	[11, 14, 16, 19, 22, 26, 28]	15	10/10	40
le450_15d	450	16,750	0.17	15	[11, 14, 16, 19, 22, 26, 28]	15	10/10	50
le450_25c	450	17,343	0.17	25	[2, 14, 19, 23, 26, 28]	25	10/10	120
le450_25d	450	17,425	0.17	25	[2, 14, 19, 23, 26, 28]	25	10/10	42
flat300_26_0	300	21,633	0.48	26	[2, 14, 19, 26, 28]	26	10/10	40
flat300_28_0	300	21,695	0.48	29	[18, 22]	30	5/10	500
flat1000_50_0	1000	245,000	0.49	50	[11, 14, 16, 19, 21, 22, 26, 28]	50	10/10	40
flat1000_60_0	1000	245,830	0.49	60	[11, 14, 16, 19, 21, 22, 26, 28]	60	10/10	45
flat1000_76_0	1000	246,708	0.49	81	[14, 21]	-	-	-
				82	[19, 23, 26, 28]	82	10/10	280
C2000.5	2000	999,836	0.50	145	[14]	-	-	-
				146	[28]	147	1/5	8000
latin_sqr_10	900	307,350	0.76	97	[26]	98	2/10	600

other state-of-the-art coloring algorithms from the literature. Our DH-GCP algorithm was programmed in Java using the multi-agent platform Jade. The program was run on a computer with a Core I5 2.5 GHz, 8 GB of RAM.

Each instance was solved 10 times independently (5 times for very large graphs). The algorithm was stopped when a legal k-coloring was found or the fixed execution timeout was reached. For all instances, a timeout limit of 10 days was used except for the large graph C2000.5 where a limit of 20 days (note that large computing times were usually allowed in the literature on GCP). We adjusted the parameters of the proposed algorithms by an experimental study. The number of iterations for each tabu search agent ($iter\_max$) was fixed to 1000. The parameters max_opt (for mediator agent) and max_opt_TS (for tabu search agent), that evaluate the improvement of solutions between generations, were fixed to 20 and 2 respectively. For $interval$, we considered the same value

10 for the same agents. The rate μ used in updating the weight matrices was fixed to 0.9.

Table 1 summarizes the computational results of our DH-GCP algorithm. Columns 2–4 show the features of the tested instance: the number of vertices ($|V|$), the number of edges ($|E|$) and the density of the graph (*dens*). Columns 5 and 6 corresponds to the best known results k^* ever reported in the literature and the corresponding references. The remaining columns give the computational results of our DH-GCP algorithm: the smallest number of colors needed to obtain a legal k-coloring, the success rate (*#hit*) and the average time for reaching the best legal k-coloring (*time* in minutes).

Table 1 shows that the results obtained by our DH-GCP algorithm are competitive with respect to many state of the art algorithms in terms of solution quality (i.e., the number of colors used). It can reach previous best known results except for 7 very difficult cases (DSJC500.5, DSJC1000.5, flat300_28_0, flat1000_76_0, latin_sqr_10, C2000.5 and R1000.5) for which few algorithms are able to attain the best known results. For these 7 instances, the deviation between our results and the best-known results is respectively 0.021 (for DSJC500.5), 0.012 (for DSJC1000.5), 0.034 (for flat300_28_0), 0.012 (for flat1000_76_0), 0.002 (for R1000.5), 0.013 (for C2000.5) and 0.01 (for latin_sqr_10) respectively. Even if we do not show detailed comparisons with individual algorithms due to space limit, we mention that the results achieved by DH-GCP remain competitive compared with many reference coloring algorithms in terms of solution quality.

4   Conclusion

The proposed distributed hybrid algorithm for the Graph Coloring Problem (DH-GCP) relies on the principles of multi-agent systems to explore a search space with the help of an ensemble of working agents (tabu search agents, crossover agents, perturbation agents). These agents are coordinated by a mediator agent using a reinforcement learning mechanism in order to make right search decisions. Decisions are influenced by a learning-based probabilistic strategy which dynamically adjusts the application probability of a particular action under a specific condition. According to whether the search process needs to be intensified or diversified, the mediator agent triggers, based on a weight matrix, either an intensification agent (tabu search agents) or a diversification agent (perturbation agents, crossover agents).

The proposed algorithm was assessed on a set of 23 difficult DIMACS coloring benchmarks. The computational results showed that DH-GCP was able to reach the previous best known results except for 7 very difficult cases and remains competitive compared to many coloring algorithms. On the other hand, the current version of the algorithm, which is a proof-of-concept prototype, is rather time consuming, partially due to the multi-agent platform Jade used for its implementation. One possible way to improve the computational efficiency of the algorithm would be to envisage a dedicated distributed implementation. Finally, both this work and the previous study on the quadratic assignment problem [25] demonstrate that the proposed framework is general enough to be adapted to solve other combinatorial search problems. It would be worthy of investigating this multi-agent based optimization framework within other settings.

Acknowledgments. We are grateful to the referees for valuable suggestions and comments which helped us improve the paper. The work is partially supported by the PGMO project (2013-2015, Jacques Hadamard Mathematical Foundation, Paris).

References

1. Avanthay, C., Hertz, A., Zufferey, N.: A variable neighborhood search for graph coloring. Eur. J. Oper. Res. **151**(2), 379–388 (2003)
2. Blochliger, I., Zufferey, N.: A graph coloring heuristic using partial solutions and a reactive Tabu scheme. Comput. Oper. Res. **35**(3), 960–975 (2008)
3. Brélaz, D.: New methods to color the vertices of a graph. Commun. ACM **22**(4), 251–256 (1979)
4. Chalupa, D.: Population-based and learning-based metaheuristic algorithms for the graph coloring problem. In: Krasnogor, N., Lanzi, P.L. (eds.) GECCO, pp. 465–472. ACM (2011)
5. Chiarandini, M., Stützle, T.: An application of iterated local search to graph coloring. In: Johnson, D.S., Mehrotra, A., Trick, M. (eds.) Proceedings of the Computational Symposium on Graph Coloring and its Generalizations, pp. 112–125, Ithaca, New York, USA (2002)
6. Dorne, R., Hao, J.K.: Tabu search for graph coloring, T-coloring and set T-colorings. In: Osman, I.H., et al. (eds.) Metaheuristics 1998: Theory and Applications, Chap. 3. Kluver Academic Publishers, Boston (1998)
7. Dorne, R., Hao, J.-K.: A new genetic local search algorithm for graph coloring. In: Eiben, A.E., Bäck, T., Schoenauer, M., Schwefel, H.-P. (eds.) PPSN 1998. LNCS, vol. 1498, pp. 745–754. Springer, Heidelberg (1998)
8. Johnson, D.S., Trick, M. (eds.): Cliques, Coloring, And Satisfiability: Second DIMACS Implementation Challenge. DIMACS Series in Discrete Mathematics and Theoretical Computer Science, vol. 26. American Mathematical Society, Boston (1996)
9. Fleurent, C., Ferland, J.A.: Genetic and hybrid algorithms for graph coloring. Ann. Oper. Res. **63**, 437–461 (1996)
10. Galinier, P., Hao, J.K.: Hybrid evolutionary algorithms for graph coloring. J. Comb. Optim. **3**(4), 379–397 (1999)
11. Galinier, P., Hertz, A., Zufferey, N.: An adaptive memory algorithm for the K-colouring problem. Discrete Appl. Math. **156**(2), 267–279 (2008)
12. Galinier, P., Hamiez, J.-P., Hao, J.-K., Porumbel, D.: Recent advances in graph vertex coloring. In: Zelinka, I., Snasel, V., Abraham, A. (eds.) Handbook of Optimization. ISRL, vol. 38, pp. 505–528. Springer, Heidelberg (2013)
13. Guo, Y., Goncalves, Y., Hsu, T.: A multi-agent based self-adaptive genetic algorithm for the long-term car pooling problem. J. Math. Model. Algorithms Oper. Res. **12**(1), 45–66 (2013)
14. Hao, J.K., Wu, Q.: Improving the extraction and expansion method for large graph coloring. Discrete Appl. Math. **160**(16–17), 2397–2407 (2012)
15. Hertz, A., de Werra, D.: Using Tabu search techniques for graph coloring. Computing **39**(4), 345–351 (1987)

16. Hertz, A., Plumettaz, M., Zufferey, N.: Variable space search for graph coloring. Discrete Appl. Math. **156**(13), 2551–2560 (2008)

17. Johnson, D., Aragon, C., McGeoch, L., Schevon, C.: Optimization by simulated annealing: an experimental evaluation; Part II, graph coloring and number partitioning. Oper. Res. **39**(3), 378–406 (1991)

18. Lü, Z., Hao, J.K.: A memetic algorithm for graph coloring. Eur. J. Oper. Res. **203**(1), 241–250 (2010)

19. Malaguti, E., Monaci, M., Toth, P.: A metaheuristic approach for the vertex coloring problem. INFORMS J. Comput. **20**(2), 302–316 (2008)

20. Malaguti, E., Toth, P.: A survey on vertex coloring problems. Int. Trans. Oper. Res. **17**(1), 1–34 (2010)

21. Moalic, L., Gondran, A.: The new memetic algorithm HEAD for graph coloring: an easy way for managing diversity. In: Ochoa, G., Chicano, F. (eds.) EvoCOP 2015. LNCS, vol. 9026, pp. 173–183. Springer, Heidelberg (2015)

22. Porumbel, D., Hao, J.K., Kuntz, P.: A search space "cartography" for guiding graph coloring heuristics. Comput. Oper. Res. **37**(4), 769–778 (2010)

23. Porumbel, D., Hao, J.K., Kuntz, P.: An evolutionary approach with diversity guarantee and well-informed grouping recombination for graph coloring. Comput. Oper. Res. **37**(10), 1822–1832 (2010)

24. Prestwich, S.: Coloration neighbourhood search with forward checking. Ann. Math. Artif. Intell. **34**(4), 327–340 (2002)

25. Sghir, I., Hao, J.K., Ben Jaafar, I., Ghédira, K.: A multi-agent based optimization method applied to the quadratic assignment problem. Expert Syst. Appl. **42**(23), 9252–9263 (2015)

26. Titiloye, O., Crispin, A.: Graph coloring with a distributed hybrid quantum annealing algorithm. In: O'Shea, J., Nguyen, N.T., Crockett, K., Howlett, R.J., Jain, L.C. (eds.) KES-AMSTA 2011. LNCS, vol. 6682, pp. 553–562. Springer, Heidelberg (2011)

27. Trick, M.A., Yildiz, H.: A large neighborhood search heuristic for graph coloring. In: Van Hentenryck, P., Wolsey, L.A. (eds.) CPAIOR 2007. LNCS, vol. 4510, pp. 346–360. Springer, Heidelberg (2007)

28. Wu, Q., Hao, J.K.: Coloring large graphs based on independent set extraction. Comput. Oper. Res. **39**, 283–290 (2012)

Variance Reduction in Population-Based Optimization: Application to Unit Commitment

Jean-Joseph Christophe, Jérémie Decock, Jialin Liu$^{(\boxtimes)}$, and Olivier Teytaud

Univ. Paris-Sud, Bat 660 Claude Shannon, 91190 Gif-sur-Yvette, France
{jean-joseph.christophe,jeremie.decock,jialin.liu,
olivier.teytaud}@inria.fr
https://tao.lri.fr

Abstract. We consider noisy optimization and some traditional variance reduction techniques aimed at improving the convergence rate, namely (i) common random numbers (CRN), which is relevant for population-based noisy optimization and (ii) stratified sampling, which is relevant for most noisy optimization problems. We present artificial models of noise for which common random numbers are very efficient, and artificial models of noise for which common random numbers are detrimental. We then experiment on a desperately expensive unit commitment problem. As expected, stratified sampling is never detrimental. Nonetheless, in practice, common random numbers provided, by far, most of the improvement.

Keywords: Noisy optimization · Variance reduction · Stratified sampling · Common random numbers

1 Introduction

1.1 Noisy Black-Box Optimization

We consider a function $f(x, w)$, with x in a d-dimensional search domain and w a random variable with values in $D \subset \mathbb{R}$. We assume that the optimization algorithm has only access to independent random realizations of $f(x, w)$. The goal of the optimization algorithm is to approximate $x^* = \arg\min_{x \in \mathbb{R}^d} \mathbb{E}_w[f(x, w)]$.

1.2 Noisy Optimization with Variance Reduction

In standard noisy optimization frameworks, the black-box noisy optimization algorithm, for its nth request to the black-box objective function, can only provide some x and receive a realization of $f(x, w_n)$. The w_n, $n \in \{1, 2, \dots\}$, are independent samples of w. The algorithm can not influence the w_n. Contrarily to this standard setting, we here assume that the algorithm can request $f(x, w_n)$ where w_n is:

- either an independent copy of w (independent of all previously used values);
- or a previously used value w_m for some $m < n$ (m is chosen by the optimization algorithm).

Due to this possibility, *paired sampling* can be applied, i.e. the same w_n can be used several times, as explained later. In addition, we assume that we have *strata*. A stratum is a subset of D. Strata have empty intersections and their union is D (i.e. they realize a partition of D). When an independent copy of w is requested, the algorithm can decide to provide it conditionally to a chosen stratum. Thanks to strata, we can apply *stratified sampling* (Sect. 1.3).

1.3   Statistics of Variance Reduction

Monte Carlo methods are the estimation of the expected value of a random variable owing to a randomly drawn sample. Typically, in our context, $\mathbb{E}[f(x, w)]$ can be estimated as a result of $f(x, w_1), f(x, w_2), \ldots, f(x, w_n)$, where the w_i are independent copies of w, $i \in \{1, \ldots, n\}$. Laws of large numbers prove, under various assumptions, the convergence of Monte Carlo estimates such as (see [2])

$$\hat{\mathbb{E}}f(x, w) = \frac{1}{n} \sum_{i=1}^{n} f(x, w_i) \to \mathbb{E}_w f(x, w). \tag{1}$$

There are also classical techniques for improving the convergence:

- *Antithetic Variates* (symmetries): ensure some regularity of the sampling by using symmetries. For example, if the random variable w has distribution invariant by symmetry w.r.t 0, then, instead of Eq. 1, we use Eq. 2, which reduces the variance:

$$\hat{\mathbb{E}}f(x, w) = \frac{1}{n} \sum_{i=1}^{n/2} (f(x, w_i) + f(x, -w_i)). \tag{2}$$

  More sophisticated antithetic variables are possible (combining several symmetries).
- *Importance Sampling*: instead of sampling w with density dP, we sample w' with density dP'. We choose w' such that the density dP' of w' is higher in parts of the domain which are critical for the estimation. However, this change of distribution introduces a bias. Therefore, when computing the average, we change the weights of individuals by the ratio of probability densities as shown in Eq. 3 - which is an unbiased estimate.

$$\hat{\mathbb{E}}f(x, w) = \frac{1}{n} \sum_{i=1}^{n} \frac{dP(w_i)}{dP'(w_i)} f(x, w_i) \tag{3}$$

- *Quasi Monte Carlo* Methods: use samples aimed at being as uniform as possible over the domain. Quasi Monte Carlo methods are widely used in integration; thanks to modern randomized Quasi Monte Carlo methods, they are usually at least as efficient as Monte Carlo and much better in favorable situations [3,13,16,24]. There are interesting (but difficult and rather "white-box") tricks for making them applicable for time-dependent random processes with many time steps [15].

- [6] proposes to generate a finite sample which approximates a random process, optimally for some metric. This method has advantages when applied in the framework of Bellman algorithms as it can provide a tree representation, mitigating the anticipativity issue. But it is hardly applicable when aiming at the convergence to the solution for the underlying random process.
- *Control Variates*: instead of estimating $\mathbb{E}f(x,w)$, we estimate $\mathbb{E}(f(x,w) - g(x,w))$, using

$$\mathbb{E}f(x,w) = \underbrace{\mathbb{E}g(x,w)}_{A} + \underbrace{\mathbb{E}\left(f(x,w) - g(x,w)\right)}_{B}.$$

This makes sense if g is a reasonable approximation of f (so that term B has a small variance) and term A can be computed quickly (e.g. if computing g is much faster than computing f or A can be computed analytically).

- *Stratified Sampling* is the case in which each w_i is randomly drawn conditionally to a stratum. We consider that the domain of w is partitioned into disjoint strata $S_1,\ldots,S_N$. N is the number of strata. The stratification function $i \mapsto s(i)$ is chosen by the algorithm and w_i is randomly drawn conditionally to $w_i \in S_{s(i)}$.

$$\hat{\mathbb{E}}f(x,w) = \sum_{i=1}^{n} \frac{P(w \in S_{s(i)})f(x,w_i)}{\text{Cardinality}\{j \in \{1,\ldots,n\}; w_j \in S_{s(i)}\}} \qquad (4)$$

- *Common Random Numbers (CRN)*, or paired comparison, refer to the case where we want to know $\mathbb{E}f(x,w)$ for several x, and use the same samples $w_1,\ldots,w_n$ for the different possible values of x.

In this paper, we focus on stratified sampling and paired sampling, in the context of optimization with arbitrary random processes. They are easy to adapt to such a context, which is not true for other methods cited above.

Stratified Sampling. Stratified sampling involves building strata and sampling in these strata.

Simultaneously Building Strata and Sampling. There are some works doing both simultaneously, i.e. build strata adaptively depending on current samples. For example, [11,18] present an iterative algorithm which stratifies a highly skewed population into a take-all stratum and a number of take-some strata. [10] improves their algorithm by taking into account the gap between the variable used for stratifying and the random value to be integrated.

A Priori Stratification. However, frequently, strata are built in an ad hoc manner depending on the application at hand. For example, an auxiliary variable $\tilde{f}(x^*, w)$ might approximate $w \mapsto f(x^*, w)$, and then strata can be defined as a partition of the $\tilde{f}(x^*, w)$. It is also convenient for visualization, as in many cases the user is interested in viewing statistics for w leading to extreme values of $f(x^*, w)$. More generally, two criteria dictate the choice of strata:

- a small variance inside each stratum, i.e. $Var_{w|S} f(x^*, w)$ small for each stratum S, is a good idea;
- interpretable strata for visualization purpose.

The sampling can be

- *proportional*, i.e. the number of samples in each stratum S is proportional to the probability $P(w \in S)$;
- or *optimal*, i.e. the number of samples in each stratum S is proportional to a product of $P(w \in S)$ and an approximation of the standard deviation $\sqrt{Var_{w|S} f(x^*, w)}$. In this case, reweighting is necessary, as in Eq. 4.

Stratified Noisy Optimization. Compared to classical stratified Monte Carlo, an additional difficulty when working in stratified noisy optimization is that x^* is unknown, so we can not easily sample $f(x^*, w)$. Also, the strata should be used for many different x; if some of them are very different, nothing guarantees that the variance $Var_{w|S} f(x, w)$ is approximately the same for each x and for x^*. As a consequence, there are few works using stratification for noisy optimization and there is, to the best of our knowledge, no work using optimal sampling for noisy optimization, although there are many works around optimal sampling. We will here focus on the simple proportional case. In some papers [12], the word "stratified" is used for *Latin Hypercube Sampling*; we do not use it in that sense in the present paper.

Common Random Numbers & Paired Sampling. Common Random Numbers (CRN), also called correlated sampling or pairing, is a simple but powerful technique for variance reduction in noisy optimization problems. Consider $x_1, x_2 \in \mathbb{R}^d$, where d is the dimension of the search domain and w_i denotes the ith independent copy of w:

$$Var \sum_{i=1}^{n} (f(x_1, w_i) - f(x_2, w'_i))$$
$$= nVar \left(f(x_1, w_1) - f(x_2, w'_1) \right)$$
$$= nVar f(x_1, w_1) + nVar f(x_2, w'_1)$$
$$- 2nCov \left(f(x_1, w_1), f(x_2, w'_1) \right).$$

If $Cov(f(x_1, w_i), f(x_2, w'_i)) > 0$, i.e. there is a positive correlation between $f(x_1, w_i)$ and $f(x_2, w'_i)$, the estimation errors are smaller. CRN is based on $w_i = w'_i$, which is usually a simple and efficient solution for correlating $f(x_1, w_i)$

and $f(x_2, w'_i)$; there are examples in which, however, this does not lead to a positive correlation. In Sect. 2.2, we will present examples in which CRN does not work.

Pairing in Artificial Intelligence. Pairing is used in different application domains related to optimization. In games, it is a common practice to compare algorithms based on their behaviors on a finite constant set of examples [8]. The cost of labelling (i.e. the cost for finding the ground truth regarding the value of a game position) is a classical reason for this. This is different from simulating against paired random realizations (because it is usually an adversarial context rather than a stochastic one), though it is also a form of pairing and is related to our framework of dynamic optimization. More generally, paired statistical tests improve the performance of stochastic optimization methods, e.g. dynamic *Random Search* [7,25] and *Differential Evolution* [20]. I has been proposed [23] to use a paired comparison-based *Interactive Differential Evolution* method with faster rates. In *Direct Policy Search*, paired noisy optimization has been proposed in [9,21,22]. Our work follows such approaches and combines them with stratified sampling. This is developed in the next section. In *Stochastic Dynamic Programming* (SDP) [1] and its dual counterpart Dual SDP [17], the classical *Sample Average Approximation* (SAA) reduces the problem to a finite set of scenarios; the same set of random seeds is used for all the optimization run. It is indeed often difficult to do better, because there are sometimes not infinitely many scenarios available. Variants of dual SDP have also been tested with increasing set of random realizations [14] or one (new, independent) random realization per iteration [19]. A key point in SDP is that one must take care of anticipativity constraints, which are usually tackled by a special structure of the random process. This is beyond the scope of this paper; we focus on direct policy search, in which this issue is far less relevant as long as we can sample infinitely many scenarios. However, our results on the compared benefits of stratified sampling and common random numbers suggest similar tests in non direct approaches using Bellman values.

2 Algorithms

2.1 Different Forms of Pairing

For each request x_n to the objective function oracle, the algorithm also provides a set $Seed_n$ of random seeds; $Seed_n = \{seed_{n,1}, \dots, seed_{n,m_n}\}$. $\mathbb{E}f(x_n, w)$ is then approximated as $\frac{1}{m_n} \sum_{i=1}^{m_n} f(x_n, seed_{n,i})$.

One can see in the literature different kinds of pairing. The simplest one is as follows: all sets of random seeds are equal for all search points evaluated during the run, i.e. $Seed_n$ is the same for all n. The drawback of this approach is that it relies on a sample average approximation: the good news is that the objective function becomes deterministic; but the approximation of the optimum is only good up to the relevance of the chosen sample and we can not guarantee convergence to the real optimum. Variants consider m_n increasing and nested sets $Seed_n$, such as $\forall(n \in \mathbb{N}^+, i \leq m_n)$, $m_{n+1} \geq m_n$ and $seed_{n,i} = seed_{n+1,i}$. A more sophisticated version is that all random seeds are equal inside an offspring, but they are changed between offspring (see discussion above). We will test this, as an intermediate step between CRN and no pairing at all. In Sect. 2.2, we explain on an illustrative example why in some cases, pairing can be detrimental. It might therefore make sense to have partial pairing. In order to have the best of both worlds, we propose in Sect. 2.3 an algorithm for switching smoothly from full pairing to no pairing at all.

2.2 Why Common Random Numbers Can Be Detrimental

The phenomenon by which common random numbers can improve convergence rates is well understood; correlating the noise between several points tends to transform the noise into a constant additive term, which has therefore less impact - a perfectly constant additive term has (for most algorithms) no impact on the run. Setting $\alpha = 1$ in Eq. 5 (below), modelizing an objective function, provides an example in which pairing totally cancels the noise.

$$f(x, w) = ||x||^2 + \alpha w' + 20(1 - \alpha)w'' \cdot x \tag{5}$$

We here explain why CRN can be detrimental on a simple illustrative example. Let us assume (toy example) that

– We evaluate an investment policy on a wind farm.
– A key parameter is the orientation of the wind turbines.
– A crucial part of the noise is the orientation of wind.
– We evaluate 30 different individuals per generation, which are 30 different policies - each individual (policy) has a dominant orientation.
– Each policy is evaluated on 50 different simulated wind events.

With CRN: If the wind orientation (which is randomized) was on average more East than it would be on expectation, then, in case of pairing (i.e. CRN), this "East orientation bias" is the same for all evaluated policies. As a consequence, the selected individuals are more East-oriented. The next iterate is therefore biased toward East-oriented.

Without CRN: Even if the wind orientation is too much East for the simulated wind events for individual 1, such a bias is unlikely to occur for all individuals. Therefore, some individuals will be selected with a East orientation bias, but others with a West orientation bias or other biases. As a conclusion, the next iterate will incur an average of many uncorrelated random biases, which is therefore less biased.

2.3 Proposed Intermediate Algorithm

We have seen that pairing can be efficient or detrimental depending on the problem. We will here propose an intermediate algorithm (Algorithm 1), somewhere in between the paired case ($g(r) = r$) and the totally unpaired case ($g(r) >> r$).

Algorithm 1. One iteration of a population-based noisy optimization algorithm with pairing.

Require: A population-based noisy optimization algorithm (in particular, rule for generating offspring)
Require: n: current iteration number
Require: $r \in \mathbb{N}^+$: a resampling rule
Require: λ: a population size
Require: $g : \mathbb{N}^+ \to \mathbb{N}^+$: a non-decreasing mapping such that $g(r) \geq r$
1: Generate λ individuals $i_1, \ldots, i_\lambda$ to be evaluated at this iteration
2: Compute the resampling number r by the resampling rule
3: Generate $P_{r,g(r)} = (w_{r,1}, \ldots, w_{r,g(r)})$ a set of $g(r)$ random seeds (we will see below different rules)
4: Each of these λ individuals is evaluated r times with r distinct random seeds randomly drawn in the family $P_{r,g(r)}$.

The $P_{r,g(r)}$ can be

– Nested, i.e. $\forall(i, r)$, $g(r) \geq i \Rightarrow w_{r,i} = w_{r+1,i}$. The $(w_{r,i})_{i \leq g(r)}$ for a fixed r are then independent.
– Independent, i.e. all the $w_{r,i}$ are randomly independently identically drawn.

SAA is equivalent to the nested case with $n \mapsto r(n)$ constant, i.e. we always use the same set of random seeds. [14] corresponds to the nested case. Classical CRN consists in $g(r) = r$ and independent sampling.

We will design, in Sect. 3, an artificial testbed which smoothly (parametrically depending on α in Eq. 5) switches

– from an ideal case for pairing (testbed in which pairing cancels the noise, as $\alpha = 1$ in Eq. 5);
– to worst case for pairing (counterexample as illustrated above, Sect. 2.2).

and which (depending on $g(\cdot)$) switches from fully paired to fully independent. We will compare stratified sampling and paired sampling on this artificial testbed. Later, we will consider a realistic application (Sect. 4).

3 Artificial Experiments

We consider a $(\mu/\mu, \lambda)$-Self-Adaptive Evolution Strategy, with $\lambda = 8d^2$, $\mu = min(2d, \lambda/4)$ and some resampling rule $r(n) = \lceil n^d \rceil$, where n is the current iteration number. We apply Algorithm 1 with $g : \mathbb{N}^+ \mapsto \mathbb{N}^+$ defined by

$$g(r) = round(r^\beta),$$

where $\beta \geq 1$ is a parameter which regulates the pairing level. When $\beta = 1$, the function evaluations are fully paired; when $\beta \to \infty$, the function evaluations are fully independent. All experiments are performed with 10000 function evaluations and are reproduced 9999 times.

3.1   Artificial Testbed for Paired Noisy Optimization

With $w = (w', w'')$, let us define

$$f(x, w) = ||x||^2 + \alpha w' + 20(1 - \alpha)w'' \cdot x \tag{6}$$

where $\cdot$ denotes the scalar product. Two different cases are considered for the random processes:

- *Continuous Case:* w' is a unidimensional standard Gaussian random variable and w'' is a d-dimensional standard Gaussian random variable.
- *Discrete Case:* w' is a Bernoulli random variable with parameter $\frac{1}{2}$ and w'' is a vector of d independent random variables equal to 1 with probability $\frac{1}{2}$ and -1 otherwise. For the stratified sampling, in case of 4 strata, we use the 2 first components of w'', which lead to 4 different cases: one for $(-1, -1)$, one for $(1, 1)$, one for $(-1, 1)$ and one for $(1, -1)$.

The motivations for this testbed are as follows:

- It is a generalization of the classical sphere function.
- The case $\alpha = 1$ is very easy for pairing (just a *Sample Average Approximation* (SAA) is enough for fast convergence as in the noise-free case - $\beta = 1$, i.e. $g(r(n)) = r(n)$, leads to canceling noise, even with resampling number $r(n) = 1$).
- The case $\alpha = 0$ is very hard for pairing; the case $\beta = 1$ (full pairing) means that the noise has the same bias for all points.
- For the discrete framework, the stratified sampling directly reduces the dimension of the noisy case: the two first components have no more noise in the stratified case.

3.2   Experimental Results

We study

$$\mathbb{E} \frac{\log ||x||^2}{\log n_e} \tag{7}$$

(the lower the better), where x is the estimate of the optimum after $n_e = 10000$ function evaluations and the optimum is 0. Experiments are reproduced 9999 times. The continuous case leads to results in Table 1. Standard deviations are ± 0.0015 for the worst cases and are not presented. Essentially, the results are:

- When α is close to 1, small β (more pairing) is better.
- When α is close to 0, large β (nearly no pairing) is better.

In the discrete case, it is easy to define pairing: we can use strata correspond to distinct values of the two first components of w''. Using the four strata corresponding to the 2 possible values of each of the two first components of w'', we get results presented in Table 2. We still see that pairing is good or bad depending on the case (sometimes leading to no convergence whereas the non-paired case converges, see row $\alpha = 0$ in dimension 5) and never brings huge improvements; whereas stratified sampling is always a good idea in our experiments.

Table 1. Efficiency (average values) of pairing (i.e. case β small) in the continuous case. Left hand side columns (β small) have more pairing than right hand side columns. Pairing is efficient for the "gentle" noise $\alpha = 1$, up to a moderate 50 % faster; but it is harmful when $\alpha = 0$ (correlated noise). Next results will investigate stratification. Bold font shows best performance and significant improvements. Positive numbers correspond to no convergence; they are never in bold. Intermediate values of β (intermediate levels of pairing) were never significantly better than others and not clearly more robust to changes in α.

α	$\beta = 1.0$ (paired)	$\beta = 1.16$	$\beta = 1.35$	$\beta = 1.57$	$\beta = 1.82$	$\beta = 2.12$	$\beta = 2.46$ ($\simeq$ unpaired)
Dimension 2 (bold for best tested algorithm)							
$\alpha = 0$	−0.07435	−0.06654	−0.07670	−0.08581	−0.09219	**−0.09603**	−0.09344
$\alpha = 0.8$	−0.34475	−0.34661	−0.35921	−0.36253	−0.36565	−0.36709	**−0.36917**
$\alpha = 1$	**−0.75048**	−0.52772	−0.50544	−0.49794	−0.49109	−0.49339	−0.49182
Dimension 3 (bold for best tested algorithm)							
$\alpha = 0$	−0.06258	−0.06373	−0.07978	−0.09489	−0.10463	−0.10931	**−0.10977**
$\alpha = 1$	**−0.47681**	−0.43320	−0.41439	−0.41004	−0.40880	−0.40202	−0.39641
Dimension 5 (bold for best tested algorithm)							
$\alpha = 0$	0.02965	0.03964	0.04409	0.04394	0.04680	0.04826	0.04823
$\alpha = 0.8$	−0.15077	−0.15977	−0.16369	−0.16687	−0.16770	−0.16793	**−0.16920**
$\alpha = 1$	**−0.23235**	−0.23188	−0.23174	−0.23125	−0.23225	−0.23232	−0.23182
Dimension 10 (bold for best tested algorithm)							
	$\beta = 1$ (paired)						$\beta = \infty$ ($\simeq$ unpaired)
$\alpha = 0$	0.097						**−0.033**
$\alpha = 0.8$	0.038						**−0.053**
$\alpha = 1.0$	**−0.057**						−0.054

4   Real World Experiments

4.1   Paired Noisy Optimization for Dynamic Problems

Paired statistical tests (e.g. Pegasus [5]) convert a stochastic optimization problem into a deterministic and easier one. Although Pegasus can cause excessive "overfitting" (specialization to the set of considered seeds) when using a fixed number of scenarios, several methods, e.g. using *Wilcoxon signed rank sum test* or changing the scenarios during learning, can reduce the "overfitting" [21,22]. *Wilcoxon signed rank sum test* pays more attention to small improvements across all scenarios rather than large changes over the return of an individual one, so that it can reduce the "overfitting" caused by a few extreme (good or bad) scenarios. [21] also shows that using an adaptive number of trials for each policy can speed-up learning in such a CRN framework. In the present work, we use new scenarios for each generation - we assume that there is no constraint on the availability of possible realizations w. Another related existing work is [9].

Table 2. Table of results (average slope as in Eq. 7; the lower, the better) depending on α (defining the problem) and β (defining the level of pairing; $\beta = 1$ means full pairing, β large means no pairing). We see that pairing can have a positive or a negative effect. We include results with stratified sampling; which are better or much better depending on the cases. Negligible standard deviations are not presented. Numbers in the stratified case are in bold when they outperform the non stratified setting.

α	$\beta = 1.0$	$\beta = 1.16$	$\beta = 1.35$	$\beta = 1.57$	$\beta = 1.82$	$\beta = 2.12$	$\beta = 2.46$
Dimension 2, no stratified sampling (bold for signif. best)							
$\alpha = 0$	−0.07200	−0.06392	−0.07926	−0.08873	−0.09539	−0.09443	−0.09382
$\alpha = 1$	−0.74716	−0.52659	−0.50665	−0.49758	−0.49383	−0.49402	−0.49310
Dimension 3, no stratified sampling (bold for signif. best)							
$\alpha = 0$	−0.00802	−0.00519	−0.01246	−0.01672	−0.01750	−0.01660	−0.01635
$\alpha = 0.4$	−0.09327	−0.10422	−0.11704	−0.12771	−0.13248	−0.13375	−0.13138
$\alpha = 0.8$	−0.25365	−0.27016	−0.28168	−0.29045	−0.29341	−0.29459	−0.29474
$\alpha = 1$	−0.39480	−0.38398	−0.37981	−0.37504	−0.37562	−0.37646	−0.37653
Dimension 3, stratified sampling (bold if better than no stratification)							
$\alpha = 0$	**−0.01931**	**−0.01396**	**−0.02585**	**−0.03590**	**−0.04430**	**−0.04836**	**−0.04744**
$\alpha = 0.8$	**−0.26548**	**−0.28079**	**−0.29481**	**−0.30133**	**−0.30797**	**−0.30761**	**−0.30763**
$\alpha = 1$	**−0.39714**	−0.38346	**−0.38021**	**−0.37749**	−0.37411	−0.37614	−0.37442
Dimension 5, no stratified sampling (bold for signif. best)							
$\alpha = 0$	0.03285	0.04253	0.04896	0.04962	0.05125	0.05336	0.05412
$\alpha = 1$	−0.23188	−0.23207	−0.23265	−0.23080	−0.23219	−0.23148	−0.23042
Dimension 5, stratified sampling (bold if better than no stratification)							
$\alpha = 0$	0.00197	−0.00880	−0.02657	−0.04158	−0.04991	**−0.05404**	−0.04617
$\alpha = 1$	**−0.23294**	−0.23146	−0.23161	**−0.23150**	**−0.23228**	**−0.23158**	**−0.23198**

	Dimension 10, no stratified sampling (bold for signif. best)						
	$\beta = 1$ (paired)						$\beta = \infty$ ($\simeq$ unpaired)
$\alpha = 0$	0.108						−0.105
$\alpha = 0.8$	0.012						−0.072
$\alpha = 1$	−0.056						−0.055
Dimension 10, stratified sampling (bold if better than no stratification)							
$\alpha = 0$	0.047						**−0.106**
$\alpha = 0.8$	**−0.033**						−0.072
$\alpha = 1$	**−0.057**						**−0.056**

It compares *Independent Random Numbers* (IRN), *Common Random Numbers* (CRN) and *Partial Common Random Numbers* (PCRN, which use pairing in the sense that the same pseudo-random numbers are used several times but in different orders) for *Simultaneous Perturbation Stochastic Approximation* and *Finite Differences Stochastic Approximation*. Both algorithms are faster when using CRN. The present work is dedicated to evolution strategies.

4.2 Unit Commitment Problem

For real world experiments, we consider the following sequential decision making problem in the *Markov Decision Processes* (MDP) framework, using discrete time steps: 10 batteries are managed to store energy bought and sold on the electricity market and 10 decision variables have to be made at each time step (i.e. the quantity of energy to buy or to sell for each battery) in order to maximize profits. We apply *rolling planning*, also known as *shrinking horizon*, i.e. new forecasts are used for updating the decisions. There are 168 time steps, i.e. 7 days with one hour per time step. We use an *operational horizon* $o = 5$ time steps, i.e. decisions are made by groups of 5 time steps. When a decision is made, it covers 5 decisions and there is no recourse on these decisions. We have a *tactical horizon* $h = 10$ time steps, i.e. we optimize over the 10 next time steps to speed up computations instead of doing it for all remaining time steps.

4.3 Testbed

We define the following variables: x is the vector of the weights of a neural network; x parametrizes the energy policy described in Eqs. 8 and 9 and d is the dimension of x. w is a random process modeling the market price. The policy (Eq. 8) uses a neural network to decide the parameters (Eq. 9) of the valorization function. The valorization function provides an estimate of the marginal value of each stock; that is, it provides, for each stock, how much (on the reward over the tactical horizon) we are willing to pay for increasing this stock by one unit.

$$d_t = \arg\max(\ reward\ over\ (t,\dots,t+h)\ ) + \sum_{i=1}^{d'} \zeta_i s_{t+h,i}. \qquad (8)$$

Each state variable corresponds to a stock. We see in Eq. 8 a compromise between the current reward (first term) and the sum $\sum_{i=1}^{10} \zeta_i \times s_{t+h,i}$ over stocks (second term). The ζ_i are estimates of the marginal values of each stock by the neural network. In Eq. 8, d_t is the vector of decisions to apply from the current time step t to time step $t+h$; $s_{t+h} = (s_{t+h,1}, \dots, s_{t+h,d'})$ is the state at the end of the tactical horizon (the quantity of energy contained in each of the 10 batteries); d' is the number of outputs of the neural network. It is equal to the number of stocks, as we have one marginal value per stock. ζ_i is the ith output of the neural network:

$$(\zeta_1, \dots, \zeta_{d'}) = neuralNetwork(x, s_t). \qquad (9)$$

$s_{t+h,i}$ depends on the random process and the decision:

$$(reward_t, s_{t+h}) = transition(s_t, d_t, random\ process). \qquad (10)$$

$reward_t$ is the reward over the operational horizon, i.e. from time t to $t+o$, i.e. $t + 5$. The *transition* function describes the problem. We use a (μ, λ)-evolution strategy to optimize x according to the objective function $f(x, w)$. $f(x, w)$ is the simulation function: it applies repeatedly the policy (Eq. 8) and the *transition* function (Eq. 10) from an initial state s_0 to a final state s_{168}. The returned value is the cumulative reward, i.e. the sum of the $reward_t$. The following setup is used: $d = 60$; $\lambda = 4(d+1) = 244$; $\mu = \lambda/4$; $r(n) = \lceil 10\sqrt{n+1} \rceil$. We define paired optimization (a.k.a common random numbers) and stratified sampling in such a case:

- We apply an evolutionary algorithm for optimizing the parameters (i.e. the weights) $x = (x_1, \ldots, x_{60})$ of the neural network controller.
- Each evaluation is a Monte Carlo average reward for a vector of parameters; a Monte Carlo evaluation is a call to $f(x, w)$ above.
- These evaluations are either pure Monte Carlo, paired Monte Carlo, stratified Monte Carlo or paired stratified Monte Carlo.

Common Random Numbers for Energy Policies: In the case of CRN (also known as pairing) for the specific case of energy policies, we apply $g(r(n)) = r(n)$, i.e. the same random outcomes $w_1, \ldots, w_{r(n)}$ are used for all individuals of a generation. The random outcomes $w_1, \ldots, w_{r(n)}$ are independently drawn for each new generation.

Stratified Sampling for Energy Policies: Stratification in the general case was defined earlier; we here discuss the application to our specific problem. It is very natural, as far as possible, to ensure that points are equally sampled among the 25 % best cases, the 25 % worst cases, the second quartile and the third quartile.

Even if these categories can only be approximately evaluated, this should decrease the variance. It is usually a good idea to stratify according to quantiles of a quantity which is as related as possible to the quantity to be averaged, i.e. $f(x, w)$. The four strata are the four quantiles on the annual average of an important scalar component of the noise.

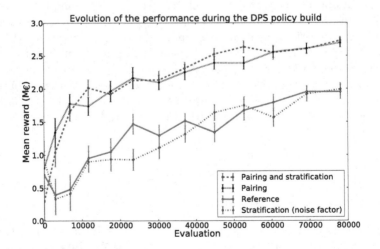

Fig. 1. X-axis: evaluation index. Y-axis: reward (the higher the better). We see that pairing is very efficient whereas stratification provides no clear improvement.

Experimental results in Fig. 1 show that pairing provides huge improvement in the realistic case. Stratification has a minor impact.

5   Conclusions

We tested, in an artificial test case and a Direct Policy Search problem in power management, paired optimization (a.k.a common random numbers) and partial variants of it. We also tested stratified sampling. Both algorithms are easy to implement, "almost" black-box and applicable for most applications. Paired optimization is unstable; it can be efficient in simple cases, but detrimental with more difficult models of noise, as shown by results on $\alpha = 1$ (positive effect) and $\alpha = 0$ (negative effect) in the artificial case (Eq. 5). We provided illustrative examples of such a detrimental effect (Sect. 2.2). Stratification had sometimes a positive effect on the artificial test case and was never detrimental. Nonetheless, on the realistic problem, pairing provided a great improvement, much more than stratification. Pairing and stratification are not totally black box; however, implementing stratification and pairing is usually easy and fast and we could do it easily on our realistic problem. We tested an intermediate algorithm with a parameter for switching smoothly from fully paired noisy optimization to totally unpaired noisy optimization. However, this parametrized algorithm (intermediate values of β) was not clearly better than the fully unpaired algorithm ($\beta = \infty$). It was not more robust in the case $\alpha = 0$, unless β is so large that there is essentially no pairing at all. As a conclusion, we firmly recommend common random numbers for population-based noisy optimization. Realistic counter-examples to CRN's efficiency would be welcome - we had such detrimental effects only in artificially built counter-example. There are probably cases (e.g. problems with rare critical cases) in which stratification also helps a lot, though this was not established in our application (which does not have natural strata).

Further Work. Other variance reduction techniques are possible. A nice challenge for future research is to find algorithms protecting variance reduction techniques from their possible detrimental effects (e.g. as efficient as CRN when $\alpha = 1$ in Eq. 5 and as efficient as no pairing when $\alpha = 0$). In particular importance sampling with optimal allocation per stratum (though we need variance estimates for that, which is difficult in a noisy optimization setting), quasi Monte Carlo (more difficult in a nearly black-box setting), or quantization [4,6].

Also, we used $g(r) = round(r^\beta)$. Results were somehow disappointing. Maybe more subtle formulas, with $g(r) = round(Ar^B)$, could be used instead, in particular $B = 1$ and $A > 1$; or $g(r)$ might be made adaptive.

References

1. Bellman, R.: Dynamic Programming. Princeton Univ, Press (1957)
2. Billingsley, P.: Probability and Measure. John Wiley and Sons, New York (1986)
3. Cranley, R., Patterson, T.: Randomization of number theoretic methods for multiple integration. SIAM J. Numer. Anal. **13**(6), 904–914 (1976)

4. Defourny, B.: Machine learning solution methods for multistage stochastic programming. Ph.D. thesis, Institut Montefiore, Université de Liège (2010)
5. Dowell, M., Jarratt, P.: The "pegasus" method for computing the root of an equation. BIT Numer. Math. **12**(4), 503–508 (1972). http://dx.org/10.1007/BF01932959
6. Dupacov, J., Gröwe-Kuska, N., Römisch, W.: Scenario reduction in stochastic programming: an approach using probability metrics. Math. Programm. **95**, 3 (2003). No. 20 in Stochastic Programming E-Print Series, Institut fr Mathematik. Springer, Berlin (2000). http://edoc.hu-berlin.de/docviews/abstract.php?id=26613
7. Hamzaçebi, C., Kutay, F.: Continuous functions minimization by dynamic random search technique. Appl. Math. Model. **31**(10), 2189–2198 (2007)
8. Huang, S.-C., Coulom, R., Lin, S.-S.: Monte-carlo simulation balancing in practice. In: van den Herik, H.J., Iida, H., Plaat, A. (eds.) CG 2010. LNCS, vol. 6515, pp. 81–92. Springer, Heidelberg (2011)
9. Kleinman, N.L., Spall, J.C., Naiman, D.Q.: Simulation-based optimization with stochastic approximation using common random numbers. Manag. Sci. **45**(11), 1570–1578 (1999). http://pubsonline.informs.org//abs/10.1287/mnsc.45.11.1570
10. Kozak, M.: Optimal stratification using random search method in agricultural surveys. Stat. Trans. **6**(5), 797–806 (2004). http://www.researchgate.net/publication/229051808_Optimal_stratification_using_random_search_method_in_agricultural_surveys/file/d912f5062bc010dd58.pdf
11. Lavallée, P., Hidiroglou, M.: On the stratification of skewed populations. Surv. Method. **14**(1), 33–43 (1988). http://www.amstat.org/sections/srms/Proceedings/papers/1987_142.pdf
12. Linderoth, J., Shapiro, A., Wright, S.: The empirical behavior of sampling methods for stochastic programming. Ann. OR **142**(1), 215–241 (2006). http://dblp.uni-trier.de/db/journals/anor/anor142.html#LinderothSW06
13. Mascagni, M., Chi, H.: On the scrambled halton sequence. Monte-Carlo Methods Appl. **10**(3), 435–442 (2004)
14. de Matos, V., Philpott, A., Finardi, E.: Improving the performance of stochastic dual dynamic programming. Applications - OR and Management Sciences (Scheduling) (2012). http://www.optimization-online.org/DB_FILE/2012/07/3529.pdf
15. Morokoff, W.J.: Generating quasi-random paths for stochastic processes. **40**(4), 765–788. http://epubs.siam.org/sam-bin/dbq/article/31795
16. Niederreiter, H.: Random number generation and quasi-monte carlo methods (1992)
17. Pereira, M.V.F., Pinto, L.: Multi-stage stochastic optimization applied to energy planning. Math. Program. 52(2), 359–375. http://dx.org/10.1007/BF01582895
18. Sethi, V.: A note on optimum stratification of populations for estimating the population means. Aust. J. Stat. **5**(1), 20–33 (1963)
19. Shapiro, A., Tekaya, W., da Costa, J.P., Soares, M.P.: Risk neutral and risk averse stochastic dual dynamic programming method. Eur. J. Oper. Res. **224**(2), 375–391 (2013)
20. Storn, R., Price, K.: Differential evolution-a simple and efficient heuristic for global optimization over continuous spaces. J. Global Optim. **11**(4), 341–359 (1997)
21. Strens, M., Lx, H.G., Moore, A., Brodley, E., Danyluk, A.: Policy search using paired comparisons. J. Mach. Learn. Res. **3**, 921–950 (2002)
22. Strens, M., Moore, A.: Direct policy search using paired statistical tests. In: Proceedings of the 18th International Conference on Machine Learning, pp. 545–552. Morgan Kaufmann, San Francisco (2001)

23. Takagi, H., Pallez, D.: Paired comparison-based interactive differential evolution. In: World Congress on Nature & Biologically Inspired Computing, NaBIC 2009, pp. 475–480. IEEE (2009)
24. Wang, X., Hickernell, F.: Randomized halton sequences. Math. Comput. Model. **32**, 887–899 (2000)
25. Zabinsky, Z.B.: Random Search Algorithms: Encyclopedia of Operations Research and Management Science. Wiley, New York (2009)

On the Codimension of the Set of Optima: Large Scale Optimisation with Few Relevant Variables

Vincent Berthier[(✉)] and Olivier Teytaud

TAO (Inria), LRI, UMR 8623 (CNRS - University of Paris-Sud), Bat 660 Claude Shannon University of Paris-Sud, 91190 Gif-sur-Yvette, France
{vincent.berthier,olivier.teytaud}@inria.fr

Abstract. The complexity of continuous optimisation by comparison-based algorithms has been developed in several recent papers. Roughly speaking, these papers conclude that a precision ϵ can be reached with cost $\Theta(n \log(1/\epsilon))$ in dimension n within polylogarithmic factors for the sphere function. Compared to other (non comparison-based) algorithms, this rate is not excellent; on the other hand, it is classically considered that comparison-based algorithms have some robustness advantages, as well as scalability on parallel machines and simplicity. In the present paper we show another advantage, namely resilience to useless variables, thanks to a complexity bound $\Theta(m \log(1/\epsilon))$ where m is the codimension of the set of optima, possibly $m << n$. In addition, experiments show that some evolutionary algorithms have a negligible computational complexity even in high dimension, making them practical for huge problems with many useless variables.

1 Introduction

In many, if not most, optimisation problems, different variables have different weight in the evaluation of the fitness function: one such example is the simple ellipsoid $f(\mathbf{x}) = 10^6 x_1^2 + \sum_{i=2}^{D} x_i^2$, where one variable ($x_1$) has a "weight" one million times more important than the other variables. We say that the condition number of the problem is of one million. In some cases though, some variables do not only have a far lesser impact on the evaluation function than others, their impact is nil. By opposition to the other "critical" variables, they are called "useless". One such case can be seen when optimising a neural network controller with a sparsity criteria where many weights are set as zero: all variables linked to neurons with those weights have no impact on the fitness function. More importantly, this phenomenon can be seen in parameter estimation problems or in genetic programming where many variables may be useless due to some other variables. Typically, many parts of a program evolved by genetic programming are not used [2] and all variables related to these parts have no impact whatsoever on the fitness function and are difficult to find [5,11,12,29,37]. In fact, [27,38] showed that removing these unused parts can be harmful. The same thing can be observed in reinforcement learning [25,33,41], evolution of trees [44], Nash equilibrium [39] or Support Vector Machines [16]. [32] also mentions very flat directions as a key point in some optimisation problems. An important question

© Springer International Publishing Switzerland 2016
S. Bonnevay et al. (Eds.): EA 2015, LNCS 9554, pp. 234–247, 2016.
DOI: 10.1007/978-3-319-31471-6_18

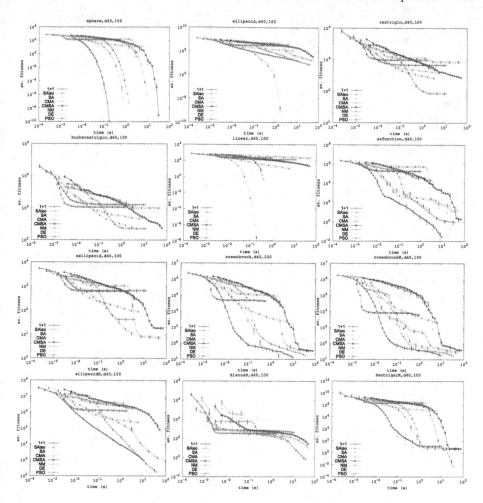

Fig. 1. Expected fitness value w.r.t computation time, for functions f1 to f12 in Bbob, respectively, in the case of 100 useless variables. A zoomable and colored version is available at http://www.lri.fr/~teytaud/uv.pdf.

is then to know how and when those useless variables impact the optimisation process, and if it is possible to overcome it.

Notations. We here introduce some notations that will be used throughout this paper. d is the dimension of the search space; we consider optimisation in $D = (0,1)^d$. m is the codimension of the set of optima, *ie.* $m = d - u$ where u is the dimension of the set of optima. x^* is an optimum of the objective function. The objective function, also known as fitness function, is $f : D \to \mathbb{R}$. $\tilde{O}$ denotes an upper bound within polylogarithmic factors.

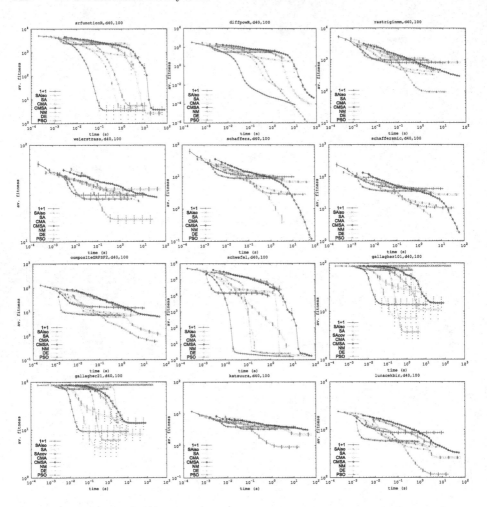

Fig. 2. Expected fitness value w.r.t computation time, for functions f13 to f24 in Bbob, respectively, in the case of 100 useless variables. A zoomable and colored version is available at http://www.lri.fr/~teytaud/uv.pdf. Confidence intervals are displayed for one point out of four; they are very small and almost invisible.

Impact of useless variables on algorithms initialisation. Some optimisers have a population size linear in the number of variables: Newuoa [32] generates an initial population of size $2d + 1$. Newuoa uses this population for building a first approximation of the Hessian. Nelder-Mead generates an initial population of size $d + 1$. Only when this initial population is generated, points which depend on the fitness values are generated based on the ranking of this initial population. Finite-differences methods will generate an initial population of size $d + 1$ for estimating the gradient. For those optimisers, we can easily see that a small number of useless variables is not an issue, but it soon becomes one

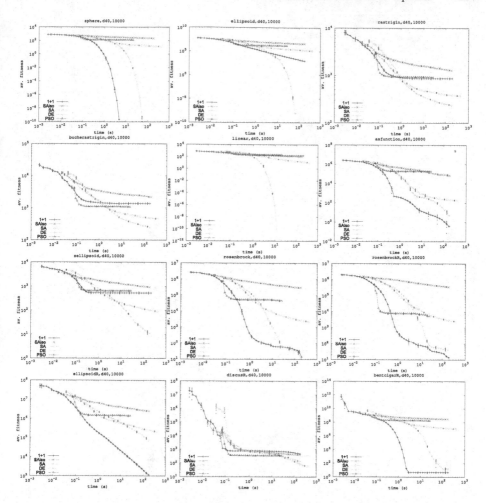

Fig. 3. Expected fitness value w.r.t computation time, for functions f1 to f12 in Bbob, respectively, in the case of 10000 useless variables. A zoomable and colored version is available at http://www.lri.fr/~teytaud/uv.pdf.

as their number increases. In practice it is often unfeasible: a population of one million individuals of one million double variables requires 16 tera-bytes of RAM (depending on double precision on the considered system). Many Evolution Strategies have a dimension-independent population size, or at worse a logarithmically increasing one. However, those that rely on covariance matrix adaptation (*eg.* CMA-ES, CMSA-ES, *etc.*) suffer from the same kind of problem: at some point, the ressources needed to store this matrix become insufficient. Other algorithms, not suffering from either of those problems, can be said to be robust w.r.t. useless variables.

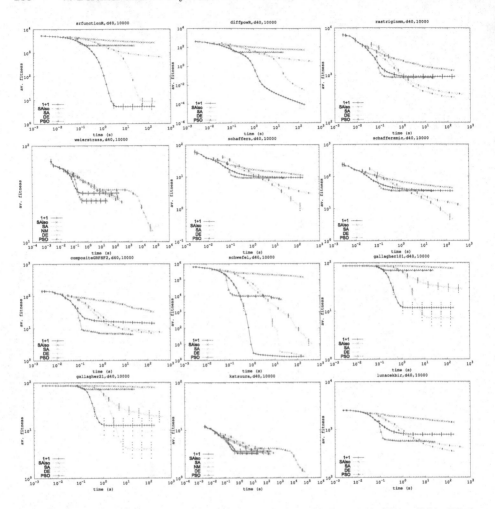

Fig. 4. Expected fitness value w.r.t computation time, for functions f13 to f24 in Bbob, respectively, in the case of 10000 useless variables. A zoomable and colored version is available at http://www.lri.fr/~teytaud/uv.pdf.

Runtimes in the presence of useless variables. When assessing the performances of an optimiser, two measures can be used. The first and arguably most used one is to compare them by the number of function evaluations required to reach the optimum. As it is independant of implementation, it is easier to use. However, there are huge gaps between the "internal costs" of different optimisation algorithms: this cost can be very high for algorithms based on covariance matrix adaptation. In fact, it can be so high that those algorithms are unable to deal with problems of dimension 10'000 or more. On the other hand, some algorithms (*eg.* Differential Evolution, Particle Swarm Optimisation, *etc.*) can be used with a hundred times more variables without problem.

The second possible measure is to compare algorithms on their runtimes: in some cases, the number of function evaluations is not important, as long as we can get the result fast. This however is a difficult measure to use: it is implementation dependent, making indirect comparisons (*eg.* from two different papers) at best suspect; it does not make any difference between the time needed to perform a function evaluation, and the time needed by the algorithm itself. In most cases, the later is supposed negligible compared to the former. With a high number of variables, this assumption does not hold anymore in some cases: CMSA-ES and CMA-ES which need to compute the eigen values and eigen vectors of the covariance matrix require a lot of time, far more than necessary for a function evaluation.

2   Theoretical Analysis: Impact of the Codimension on the Required Number of Function Evaluations

We first summarize the state of the art. We then study lower bounds (Sect. 2.1) and upper bounds (Sect. 2.2). We first discuss the case of a codimension m equal to the dimension d, *ie.* the set of optima has dimension 0 - for example a single optimum. Sections 2.1 and 2.2 will discuss the extension of these results to codimension $m < d$. [14] has shown that the number of function comparisons for finding the optimum with precision ϵ is $\Theta(d \log(1/\epsilon))$ for algorithms based on comparisons. The upper bound is for some specific comparison-based algorithm on the sphere function and the lower bound is in the case of any family of functions with unique optimum, when the optimum can be anywhere in the domain (optimum uniformly randomly drawn in the domain, or worst case over optima in the domain), and for a precision (stopping criterion) defined either in terms of distance to the optimum, or in terms of fitness values, if the fitness values $f(x) - f(x^*) = \Omega(||x - x^*||^\alpha)$ for some $\alpha > 0$. These results are based on information theory. Basically, a comparison provides one bit of information, so if we need a precision such that the optimum should be described with M digits (in binary), we need M comparisons. More generally, a ranking of λ offspring provides at most $\log_2(\lambda!)$ bits of information, and detailed results for algorithms using a selection operator of μ individuals over λ can be derived in a similar manner. [24] obtained a more general result (including various models of noise), at the expense of a different dependency in ϵ; they get: (i) a lower bound on the number of comparisons $\Omega(d \log(1/\epsilon))$ on the number of iterations before reaching an expected precision ϵ. (ii) an upper bound on the number of comparisons $O(d \log(1/\epsilon)^2)$ on the number of comparisons before reaching an expected precision ϵ, reached by an explicit algorithm.

2.1   Lower Bound

The lower bound in [14] can be adapted to our setting as follows:

Theorem 1 (corollary of [14]**):** *Consider a fixed $\delta < 1$. Consider the function* $f_{x^*,R,d,m} : x \mapsto \sum_{i=1}^{m}(R(x - x^*))_i^2$ *where R is a rotation of $\mathbb{R}^d$ and $x^* \in D$.*

Consider F_m the set of such functions. Consider a comparison-based algorithm A. Then, there is a universal constant K (depending on δ only), such that if for all functions in F_m, with probability at least $1 - \delta$, A outputs $\hat{x}$ such that $||\hat{x} - x^*|| \leq \epsilon$ after n comparisons, then

$$n \geq K \times m \times \log(1/\epsilon).$$

Proof. Consider $F'_{m,d}$ the restriction of $F_{m,d}$ to the identity matrix for R. Consider the optimisation in $(0,1)^d \times \{0\}^{d-m}$. Then by [14], the number n ensuring precision ϵ is at least $K \times m \times \log(1/\epsilon)$, for some universal K depending on δ only. $F'_m \subset F_m$, hence a lower bound for F'_m also holds for F_m. This yields the expected result. $\qquad\square$

2.2   Upper Bound

The result from [24], for the upper bound and in the noise-free case, is as follows:

Theorem 2. (corollary of [24]**):** *Consider a fixed $\delta < 1$. Consider the function $f_{x^*,R,d,m} : x \mapsto \sum_{i=1}^{m}(R(x - x^*))_i^2$ where R is a rotation of $\mathbb{R}^d$ and $x^* \in D$. Consider $F_{m,d}$ the set of such functions, for a given d and a given m. Then, there is a universal constant K (depending on δ only) and an optimisation algorithm A, such that for all functions in $F_{m,d}$, with probability at least $1 - \delta$, A outputs $\hat{x}$ such that $||\hat{x} - x^*|| \leq \epsilon$ after n comparisons, where*

$$n = \lceil K \times m \times \tilde{O}(\log(1/\epsilon)^2) \rceil. \tag{1}$$

Proof : The algorithm in [24] uses coordinate-wise line search, which can not be applied directly for our rotated framework. However, as pointed out in [24] (Sect. 5.1: " an analysis with the same result can be obtained with [...] chosen uniformly from the unit sphere"), the same result holds with randomly rotated search directions. The algorithm with randomly rotated search direction applied to $f_{x^*,R,d,m}$ exactly mimics the behavior of the algorithm on $f_{x^*,R,m,0}$. This yields the expected result. $\qquad\square$

We point out that evolution strategies (usually) also have this invariance property. However, we did not use evolution strategies in the proof because there is no formal proof of convergence of evolution strategies. Nonetheless, [1] is close to such a result for evolution strategies (up to the sign of the constant), Theorem 2 shows an upper bound for comparison-based methods, and there is a big hope that Theorem 2 could be adapted to evolution strategies if the constant in [1] is proved negative.

The gap with the lower bound is the exponent 2 on $\log(\frac{1}{\epsilon})$ in Eq. 1. We do not reduce the gap in the general case, but we propose the following partial result, using $F''_{m,d} = \{f_{x^*,R,d,m}; \forall i \, x_i^* \neq 0, R \text{ has all coefficients in } \{0,1\}\}$.

Theorem 3 : *Consider a fixed $\delta < 1$. Consider the family $F''_{m,d}$ of objective functions. Then, there is a universal constant K (depending on δ only) and an optimisation algorithm A, using the parameter m as input, such that for all functions in $F_{m,d}$, with probability at least $1-\delta$, for ϵ sufficiently small, A outputs $\hat{x}$ such that $||\hat{x} - x^*|| \le \epsilon$ after n comparisons, where $n = \lceil K \times m \times O(\log(1/\epsilon)) \rceil$.*

Remark 1 : We prove the upper bound for permutations of coordinates, and not for the complete set of rotations. We assume that m is known; we conjecture that this assumption can be removed. The result is for ϵ sufficiently small.

Proof:

Step 1: consider many algorithms. Consider $I = \{(i_1, \ldots, i_m) \in \{1, \ldots, d\}^m; i_1 < i_2 < \cdots < i_m\}$. The cardinal of I is $z = d!/(m!(d-m)!)$. For each i, consider the algorithm A_i realizing the upper bound in [14] with probability $1 - \delta/(3z)$, for some number of function evaluations w, for any sphere function restricted to m components $i = (i_1, \ldots, i_m)$. By union bound, all the algorithms reach this bound, with probability at least $1 - \delta/3$.

Step 2: a portfolio of algorithms, and algorithm selection. Consider now the algorithm A running all the A_i concurrently, in a round. However, the algorithm spends half his computational effort on the A_i which has found the best point up to now, and distributes the remaining computational power evenly over the other A_j. So a round of A is as follows:

(i) Spend one function evaluation on each A_j, $j \in I$. This costs z function evaluations.

(ii) Spend z function evaluations on the A_{j^*}, with j^* the index of the algorithm which has proposed the best search point (randomly break ties).

The overall algorithm repeats (i) and (ii) up to the available budget.

Step 3: eventually, only the right algorithm is selected. Consider the solver A_{k^*} where k^* is the family of the $R^{-1}(e_j)$ for $j \le m$. Only this solver, among the A_j, can converge to the optimum. Hence, for ϵ' sufficiently small, A_{k^*} always wins the comparison after it reaches optimality within precision ϵ'. The upper bound states that such a precision is reached with probability at least $1 - \delta/3$ when the number of rounds is at least w.

When this precision ϵ' is reached, $j^* = k^*$, and from now on A_{k^*} spends half of the computation budget.

Step 4: the budget. We have seen that A^{k^*} spends half of the computation budget, except possibly for the early rounds (before reaching precision ϵ', see step 3). Let us now show that A^{k^*} spends one fourth of the whole computation budget, when the requested precision is small enough.

Let us choose $\epsilon < \epsilon'$ such that the required number of rounds for A_{k^*} to reach precision ϵ with probability at least $1 - \delta/3$ is at least twice more (i.e. $2w$) than the budget w. Such an ϵ exists by the lower bound. With probability $1 - \delta$, when A_{k^*} reaches such a precision ϵ,

– the overall number of rounds is at least $2w$ (by the use of the lower bound, above);

– and during the second half of these $\geq 2w$ rounds at least one half of the evaluations have been spent for A_{k^*} (by Step 3).

Therefore it has spent at least one fourth of the budget when this number of rounds $2w$ is reached.

Step 5: concluding. With probability at least $1-2\delta/3$, one fourth of the budget has been devoted to A_{k^*} when the number of rounds is $\geq 2w$. With probability at least $1 - \delta/3$, A_{k^*} has the rate provided by the upper bound. This provides the expected result.   □

3   Algorithms and Their Invariances

Section 3.1 discusses invariance in optimisation algorithms. Section 3.2 presents the optimisation algorithms we consider.

3.1   Old and New Invariances

Invariance is a classical consideration in optimisation. Let us distinguish several kinds of invariance (the fifth one is a new kind of invariance in which we are particularly interested in the present paper): (i) Invariance w.r.t. translations is hard to achieve, due to the initialisation; a probability distribution for the initial search point(s) can not be translation invariant. However, up to the initialisation issue, many algorithms are invariant by translations of the objective function. It is sufficient to prove lower bounds for evolution strategies [22,23]. (ii) Invariance by composition with increasing functions is at the heart of extensions of these lower bounds to a more general setting, using information theory [14] - basically, a comparison can provide only one bit of information, hence there is a limited rate for comparison-based algorithms. (iii) Invariance w.r.t. rotations does not always hold, as discussed below for various algorithms. Most algorithms are invariant w.r.t. permutations of indices. Anisotropic evolution strategies [3] provide invariance w.r.t. rescaling of variables (up to the initialisation), but not w.r.t rotations. (iv) Invariance w.r.t. linear transformation (not only rotations) is addressed in e.g. the Newton method in mathematical programming. It is approximated without expensive computation of the Hessian in the BFGS [8,13,17,35] method. Up to the initialisation, black-box counterparts of the quasi-Newton methods ensure similar invariances [32]. In the field of evolution strategies, the most well known methods which ensure invariance w.r.t. linear transformations are CMAES [19] and CMSA [4] both providing invariance with respect to rotations.

(v) This paper discusses another kind of invariance: the fact that an algorithm is invariant w.r.t. addition of useless variables. An algorithm is said to be invariant w.r.t. addition of useless variables if this addition has no impact on the performances of the algorithm: the best obtained fitness with and without useless variable is the same, and it is reached after the same number of function evaluations.

3.2   Algorithms Used in Our Experiments, and Their Invariances

Parameters used for the nine algorithms in our comparison are (with d as the dimension presented below.

The optimisation algorithm classically associated to our chosen testbed, namely BBOB, is CMAES [19]. We use population size $\lambda = 4 + 3\log(d)$, parent population size $\mu = \lambda/2$. CMAES has some invariance properties w.r.t. rotations and translations [20], except (as most algorithms) for the initialisation which, as discussed above, can not be translation invariant. CMAES is asymptotically invariant by rescaling of variables. On the other hand, CMAES is not invariant by addition of useless variables.

We use a Self-adaptive evolution strategy, SA [3]. It uses isotropic mutations, with population size $\lambda = 12$ and parent population size $\mu = 3$. The mutation rate for step-sizes is $\tau = 1/\sqrt{2d}$. We also consider an anisotropic variation of SA [3], with the same parameters and an added step-size mutation rate for each variable $\tau_{local} = \frac{1+d(d+1)}{6}$. It is not invariant by rotation. It is invariant for rescaling of variables, up to the initialisation. We also use Covariance Matrix Self-Adaptation, CMSA [4] with the same configuration as anisotropic SA-ES, and a learning rate for the covariance matrix $\tau_C = \frac{1+d(d+1)}{2\mu}$. CMSA has the same kind of invariances as CMAES. CMSA is the extension of SA for invariance w.r.t. rotations. The computational cost of CMSA is higher than the one of SA. As CMAES, it is not invariant by addition of useless variables. Another form of covariance learning was proposed in [34], SA with covariance. The configuration is the same as anisotropic SA-ES, with an added parameter $\beta = 0.0873$ for covariance matrix update. Invariant for all invariance criteria discussed here. Yet another algorithm is Differential evolution DE [40]; we use population size 30, DE/Curr-to-best/1, $Cr = .5$, $F_1 = F_2 = .8$. DE and combinations of DE algorithms won many competitions in evolutionary computation [10]. Invariant for all invariance criteria discussed here when $Cr = 1$; but not w.r.t rotations when $Cr < 1$. We also use the old and efficient one plus one evolution strategy with one-fifth success rate, $(1 + 1)$-ES [34], step-size multiplied by 1.5 in case of success and divided by $1.5^{0.25}$ otherwise. Invariant for all invariance criterium discussed here. Nelder-Mead [30], which has the same kind of invariances w.r.t. rotations and translations as CMAES and CMSA. Its parameters are $\alpha = 1$, $\gamma = 2$, $rho = -0.5$ and $\sigma = 0.5$. Finally, we use Particle Swarm Optimisation PSO [26,36]. We use a population size 30, a social neighbourhood of size 10, $\omega = 1/2\log(2)$, $\phi_g = \phi_p = \frac{1}{2} + \log(2)$, initial velocity $\frac{3}{4}$ and maximum velocity $\frac{3}{2}$. This parametrization is a compromise between some works for defining a standard PSO [6,9,43]. PSO is not invariant for rotations [21]. For all algorithms, the initialisation is as follows. Each coordinate of each individual is randomly drawn according to a Gaussian random variable with zero mean and standard deviation 6.

4   Experiments

Test cases & criteria. We use the functions from the BBOB test set, and perform experiments with additional useless variables, *ie.* we have codimension $m = 40$, and dimension $d = m + u$ with $u = 100, 1000000$ useless variables.

Other experiments have been performed with $m = 2, 3, 4, 5, 8, 10, 16, 20, 32, 64,$ and also with $u = 10000$; results were in agreement with results presented below with $m = 40$ and $u \in \{100, 10000\}$. We consider the expected fitness value (y-axis; the fitness at the optimum is substracted as all our algorithms are invariant by addition of a constant to the optimum), for given computation times (x-axis). The x-axis is computation time, because for large number of variables the internal computation time of considered algorithms is not negligible. In fact, many algorithms could not run at all with such high dimension. We did not permute coordinates, so that the useless variables are always the last ones. However, all considered algorithms are invariant by permutation of variables, so that this is not an issue.

Results. In all results, confidence intervals are presented for one point out of four; they are almost invisible because they are very small. Results are presented in Figs. 1 and 2 for 100 useless variables, and in Figs. 3 and 4 for 10000 useless variables. Roughly speaking, many algorithms can compete for dimension 140 (codimension 40, 100 useless variables), though the simple $(1 + 1)$-ES and DE perform best overall (recall that we consider time on the x-axis, and not the number of evaluations). With 10000 useless variables, only fast algorithms (DE/SA/SAiso) can compete; DE performs best in case of ill-conditioning; SA performs well in case of ill-conditioning and no rotation. Algorithms which are not presented in the comparison are those who could not provide results in the given time limit.

5   Conclusion

This paper emphasises useless variables as a key for understanding the practical behavior of evolutionary algorithms on high dimensional problems. On the theoretical side, we extend known runtime analysis from the case of a set of optima with dimension 0 to a set of optima with dimension > 0, leading to a codimension m possibly much lower than the dimension d. The lower bound extends the known lower bound, from *dimension = codimension* to more general cases. The upper bound holds for permutation of coordinates and not for the whole family of rotations (Theorem 3), or, in the case of full rotations, with a quadratic dependency in the log-precision (Theorem 2). Pratically speaking, whereas many methods rely on a linear number of function evaluations (typically just for the initialisation), evolutionary algorithms use a logarithmic or constant initial population size. In addition, an algorithm such as DE or SA or SAaniso or the simple $(1 + 1)$-ES will just ignore unimportant variables and optimize the remaining ones. Therefore, evolutionary algorithms can handle very large problems, provided that the problem has a a special structure - in particular, when many variables are useless; and this is far from being trivial as some state of the art optimisation methods such as Newuoa, CMAES or CMSA can not do that. In fact, a more general case might be true - when, up to a rotation, many variables are useless; in particular, DE is invariant by rotation when cross-over is disabled (*ie.* Cr=0), and (1+1)-ES is invariant by rotation, so that rotations of problems with many useless variables can be tackled. Importantly, rotations of problems with useless variables are not problems with useless variables - therefore, our results show that some high-dimensional problems can be tackled whenever they have no useless variables, but are rotations of problems with useless variables. Experimentally, we successfully optimized BBOB functions with up to a million of useless variables. Unsurprisingly, for algorithms which are invariant w.r.t. useless variables, the best fitness for a given number of evaluations is exactly the same as with no useless variables. On the other hand, results become worse for algorithms which do not have this invariance and can even become impossible to obtain in a timely fashion due to computation time constraints.

Further Work. (a) On the mathematical side, we conjecture that Theorem 3 also holds with F_m instead of F_m'', *ie.* with full rotations and not only with permutations of coordinates. (b) On the experimental side, we might study the same question empirically: what happens with random rotations of the BBOB testbed embedded in a large set of useless coordinates. For algorithms which are invariant per rotation (not DE, not PSO) this does not make any difference. (c) Adaptive methods for choosing parameters might be tested for PSO or DE [7, 28,31,42] as they could maybe handle better the extreme size of our problems. (d) We tested the addition of completely useless variables. In fact, since full separability and fully rotated problems are extreme cases, we might consider variables with very low but not zero impact. We might use tricks similar to those used in the Cute testbed for partial separability [18]. (e) Recently, an effort has been made for developing real world test functions in the evolutionary computation community [15]. This provides an example of test case in which the real world decided the level of separability and the level of useless variables in a test case. Extended [15] to a high dimension case might be a good experiment.

References

1. Auger, A.: Convergence results for $(1, \lambda)$-SA-ES using the theory of φ-irreducible Markov chains. Theor. Comput. Sci. **334**, 35–69 (2005)
2. Banzhaf, W., Langdon, W.B.: Some considerations on the reason for bloat. Genet. Program. Evolvable Mach. **3**(1), 81–91 (2002)
3. Beyer, H.G.: The Theory of Evolution Strategies. Natural Computing Series. Springer, Heideberg (2001)
4. Beyer, H.-G., Sendhoff, B.: Covariance matrix adaptation revisited – the CMSA evolution strategy –. In: Rudolph, G., Jansen, T., Lucas, S., Poloni, C., Beume, N. (eds.) PPSN 2008. LNCS, vol. 5199, pp. 123–132. Springer, Heidelberg (2008)
5. Bleuler, S., Brack, M., Thiele, L., Zitzler, E.: Multiobjective genetic programming: reducing bloat using SPEA2. In: Proceedings of the 2001 Congress on Evolutionary Computation CEC2001, pp. 536–543. IEEE Press, COEX, World Trade Center, 159 Samseong-dong, Gangnam-gu, Seoul, Korea (27–30 2001). http://citeseer.ist.psu.edu/bleuler01multiobjective.html
6. Bratton, D., Kennedy, J.: Defining a standard for particle swarm optimization. In: IEEE Swarm Intelligence Symposium, pp. 120–127 (2007). http://dx.org/10.1109/SIS.2007.368035

7. Brest, J., Greiner, S., Boskovic, B., Mernik, M., Zumer, V.: Self-adapting control parameters in differential evolution: a comparative study on numerical benchmark problems. IEEE Trans. Evol. Comput. **10**(6), 646–657 (2006)

8. Broyden, C.G.: The convergence of a class of double-rank minimization algorithms 2. New Algorithm. J. Inst. Math. Appl. **6**, 222–231 (1970)

9. Clerc, M.: Beyond standard particle swarm optimisation. IJSIR **1**(4), 46–61 (2010). http://dblp.uni-trier.de/db/journals/ijsir/ijsir1.html#Clerc10

10. Das, S., Suganthan, P.N.: Differential evolution: a survey of the state-of-the-art. IEEE Trans. Evol. Comput. **15**(1), 4–31 (2011)

11. De Jong, E.D., Watson, R.A., Pollack, J.B.: Reducing bloat and promoting diversity using multi-objective methods. In: Proceedings of the Genetic and Evolutionary Computation Conference, GECCO-2001, pp. 11–18. Morgan Kaufmann Publishers, San Francisco, CA (2001). http://citeseer.ist.psu.edu/dejong01reducing.html

12. Ekárt, A., Németh, S.Z.: Maintaining the diversity of genetic programs. In: Foster, J.A., Lutton, E., Miller, J., Ryan, C., Tettamanzi, A.G.B. (eds.) EuroGP 2002. LNCS, vol. 2278, pp. 162–171. Springer, Heidelberg (2002)

13. Fletcher, R.: A new approach to variable-metric algorithms. Comput. J. **13**, 317–322 (1970)

14. Fournier, H., Teytaud, O.: Lower bounds for comparison based evolution strategies using VC-dimension and sign patterns. Algorithmica **59**(3), 387–408 (2010). http://hal.inria.fr/inria-00452791

15. Gallagher, M.: Clustering problems for more useful benchmarking of optimization algorithms. In: Dick, G., Browne, W.N., Whigham, P., Zhang, M., Bui, L.T., Ishibuchi, H., Jin, Y., Li, X., Shi, Y., Singh, P., Tan, K.C., Tang, K. (eds.) SEAL 2014. LNCS, vol. 8886, pp. 131–142. Springer, Heidelberg (2014)

16. Girosi, F.: An equivalence between sparse approximation and support vector machines. In: Proceedings of NIpPS 10, pp. 1455–1480. Morgan Kaufmann (1998)

17. Goldfarb, D.: A family of variable-metric algorithms derived by variational means. Math. Comput. **24**, 23–26 (1970)

18. Gould, N.I.M., Orban, D., Toint, P.L.: Cuter and sifdec: a constrained and unconstrained testing environment, revisited. ACM Trans. Math. Softw. **29**(4), 373–394 (2003)

19. Hansen, N., Ostermeier, A.: Completely derandomized self-adaptation in evolution strategies. Evol. Comput. **11**(1), 159–195 (2003)

20. Hansen, N.: Adaptive Encoding for Optimization. Research report RR-6518, INRIA (2008). http://hal.inria.fr/inria-00275983/en/

21. Hansen, N., Ros, R., Mauny, N., Schoenauer, M., Auger, A.: PSO Facing Non-Separable and Ill-Conditioned Problems. Research report RR-6447, INRIA (2008). http://hal.inria.fr/inria-00250078/en/

22. Jagerskupper, J.: In between progress rate and stochastic convergence. Dagstuhl's seminar (2006)

23. Jagerskupper, J., Witt, C.: Runtime analysis of a (mu+1)es for the sphere function. Technical report (2005)

24. Jamieson, K.G., Nowak, R.D., Recht, B.: Query complexity of derivative-free optimization. In: NIPS, pp. 2681–2689 (2012)

25. Kearns, M., Mansour, Y., Ng, A.: A sparse sampling algorithm for near-optimal planning in large markov decision processes. In: IJCAI, pp. 1324–1231 (1999). http://citeseer.ist.psu.edu/kearns99sparse.html

26. Kennedy, J., Eberhart, R.C.: Particle swarm optimization. In: Proceedings of the IEEE International Conference on Neural Networks, pp. 1942–1948 (1995)

27. Langdon, W.B., Poli, R.: Fitness causes bloat: mutation. In: Koza, J. (ed.) Late Breaking Papers at GP 1997, pp. 132–140. Stanford Bookstore, Stanford (1997)
28. Liu, J., Lampinen, J.: A fuzzy adaptive differential evolution algorithm. Soft Comput. **9**(6), 448–462 (2005)
29. Luke, S., Panait, L.: A comparison of bloat control methods for genetic programming. Evol. Comput. **14**(3), 309–344 (2006)
30. Nelder, J., Mead, R.: A simplex method for function minimization. Comput. J. **7**, 308–311 (1965)
31. Pošík, P., Klemš, V.: JADE, an adaptive differential evolution algorithm, benchmarked on the BBOB noiseless testbed. In: Proceedings of the 14th Annual Conference Companion on Genetic and Evolutionary Computation, pp. 197–204. ACM (2012)
32. Powell, M.J.D.: Developments of newuoa for minimization without derivatives. IMA J. Numer. Anal., p. 649. http://dx.org/10.1093/imanum/drm047
33. Ratitch, B., Precup, D.: Sparse distributed memories for on-line value-based reinforcement learning. In: Boulicaut, J.-F., Esposito, F., Giannotti, F., Pedreschi, D. (eds.) ECML 2004. LNCS (LNAI), vol. 3201, pp. 347–358. Springer, Heidelberg (2004)
34. Rechenberg, I.: Evolutionstrategie: Optimierung Technischer Systeme nach Prinzipien des Biologischen Evolution. Fromman-Holzboog Verlag, Stuttgart (1973)
35. Shanno, D.F.: Conditioning of quasi-newton methods for function minimization. Math. Comput. **24**, 647–656 (1970)
36. Shi, Y., Eberhart, R.C.: A modified particle swarm optimizer. In: Proceedings of IEEE International Conference on Evolutionary Computation, pp. 69–73. IEEE Computer Society, Washington, DC, May 1998
37. Silva, S., Costa, E.: Dynamic limits for bloat control. In: Deb, K., Tari, Z. (eds.) GECCO 2004. LNCS, vol. 3103, pp. 666–677. Springer, Heidelberg (2004)
38. Soule, T.: Exons and code growth in genetic programming. In: Foster, J.A., Lutton, E., Miller, J., Ryan, C., Tettamanzi, A.G.B. (eds.) EuroGP 2002. LNCS, vol. 2278, pp. 142–151. Springer, Heidelberg (2002)
39. St-Pierre, D.L., Louveaux, Q., Teytaud, O.: Online sparse bandit for card games. In: van den Herik, H.J., Plaat, A. (eds.) ACG 2011. LNCS, vol. 7168, pp. 295–305. Springer, Heidelberg (2012)
40. Storn, R., Price, K.: Differential evolution: a simple and efficient heuristic for global optimization over continuous spaces. J. Global Optim. **11**(4), 341–359 (1997). http://dx.org/10.1023/A:1008202821328
41. Sutton, R.: Generalization in reinforcement learning: successful examples using sparse coarse coding. In: Touretzky, D.S., Mozer, M.C., Hasselmo, M.E. (eds.) Advancesin Neural Information Processing Systems, vol. 8, pp. 1038–1044. The MIT Press, Cambridge (1996). http://citeseer.ist.psu.edu/sutton96generalization.html
42. Yu, W.J., Zhang, J.: Multi-population differential evolution with adaptive parameter control for global optimization. In: Proceedings of the 13th Annual Conference on Genetic and Evolutionary Computation. GECCO 2011, NY, USA, pp. 1093–1098 (2011). http://doi.acm.org/10.1145/2001576.2001724
43. Zambrano-Bigiarini, M., Clerc, M., Rojas, R.: Standard particle swarm optimisation 2011 at cec-2013: a baseline for future PSO improvements. In: IEEE Congress on Evolutionary Computation, pp. 2337–2344. IEEE (2013). http://dblp.uni-trier.de/db/conf/cec/cec2013.html#Zambrano-BigiariniCR13
44. Zhang, B.T., Ohm, P., Mühlenbein, H.: Evolutionary induction of sparse neural trees. Evol. Comput. **5**(2), 213–236 (1997)

Author Index

Printed in the United States
By Bookmasters